Iodinated density gradient media

a practical approach

Edited by
D Rickwood
Department of Biology, University of Essex,
Colchester, Essex, England

ISBN 0-904147-51-7

IRL PRESS
Oxford · Washington DC

IRL Press Limited,
P.O. Box 1,
Eynsham,
Oxford OX8 1JJ,
England

British Library Cataloguing in Publication Data
Iodinated density – gradient media. –
(Practical approaches in biochemistry)
1. Biological chemistry – Technique
I. Rickwood,D. II. Series
574.19′2′028 QP519.7

ISBN 0-904147-51-7

Cover Photograph
The buoyant density of biological particles in iodinated density-gradient media

Printed in England by Information Printing Limited, Eynsham.

Preface

The publication of this book coincides with the tenth anniversary of the introduction of the first nonionic iodinated density-gradient medium, metrizamide. Over the last decade it has become apparent that, as centrifugation media, iodinated compounds have a very wide range of applications; particles from macromolecules to whole cells have been separated using these media. This book brings together all of the various techniques that have been developed using these gradient media and gives the practical details of each method. It contains some data that have already appeared in the scientific literature together with a significant amount of new data which have not been published elsewhere. Hence this book provides a unique reference guide to the use of iodinated gradient media for biological separations.

D. Rickwood

Contributors

D. Bailey
Department of Surgery, St. George's Hospital Medical School, Cranmer Terrace, London, SW17 ORE, U.K.

T. Berg
Institute for Nutrition Research, University of Oslo, P.O. Box 1046, Blindern, Oslo 3, Norway.

R. Blomhoff
Institute for Nutrition Research, University of Oslo, P.O. Box 1046, Blindern, Oslo 3, Norway.

A. Bøyum
Norwegian Defence Research Establishment, Division for Toxicology, N-2007 Kjeller, Norway

T.C. Ford
Department of Biology, University of Essex, Wivenhoe Park, Colchester, Essex, CO3 3SQ, U.K.

J. Graham
Department of Biochemistry, St. George's Hospital Medical School, Cranmer Terrace, London, SW17 ORE, U.K.

J.F. Houssais
Institut Curie, Section de Biologie, Laboratoires 110-111-112, Centre Universitaire, 91405 Orsay-Cedex, France.

H. Osmundsen
Department of Biochemistry, Veterinary College, Oslo 1, Norway.

K. Patel
Department of Biochemistry, St. George's Hospital Medical School, Cranmer Terrace, London, SW17 ORE, U.K.

D. Rickwood
Department of Biology, University of Essex, Wivenhoe Park, Colchester, Essex, CO3 3SQ, U.K.

D.A. Vanden Berghe
Laboratory of Microbiology, University of Antwerp, U.I.A., B-2610 Wilrijk, Belgium.

S. Wagner
Department of Biochemistry, St. George's Hospital Medical School, Cranmer Terrace, London, SW17 ORE, U.K.

J. Wall
Department of Surgery, St. George's Hospital Medical School, Cranmer Terrace, London, SW17 ORE, U.K.

R. Wattiaux
Laboratoire de Chimie Physiologique, Facultés Universitaires Notre-Dame de la Paix, 61, rue de Bruxelles, B-5000 Namur, Belgium.

S. Wattiaux-De Coninck
Laboratoire de Chimie Physiologique, Facultés Universitaires Notre-Dame de la Paix, 61 rue de Bruxelles, B-5000 Namur, Belgium.

Contents

CHAPTER 1

Properties of Iodinated Density-gradient Media

D. Rickwood

1. INTRODUCTION

Density-gradient separations can be carried out on the basis of size or density of particles; these two methods are termed rate-zonal and isopycnic fractionations, respectively. In the case of rate-zonal separations the primary function of the gradient is simply to stabilise the liquid column in the tube against movement resulting from convection currents (1) although a secondary role of the gradient is to produce a gradient of viscosity which helps to improve the resolution of gradients (2). For isopycnic separations the most important feature of the gradient medium is that the maximum density of its solutions is high enough to be greater than that of the particles; in this context it is important to realise that, as a result of different levels of hydration of particles in different gradient media, the densities of particles vary depending on the particular gradient medium used. Besides these primary requirements for properties of gradient media for rate-zonal and isopycnic separations, a number of other general properties of an ideal density-gradient medium have been defined (1,3) and these can be listed as follows.

(i) The compound should be inert or at least non-toxic to biological material.
(ii) The physico-chemical properties of the solutions of gradient medium should be known and it should be possible to use one or more of these properties (e.g., refractive index) to determine the precise concentration of the gradient medium.
(iii) Solutions of the gradient medium should not interfere with monitoring of the zones of fractionated material within the gradient.
(iv) It should be easy to separate the sample material from the gradient medium without loss of the sample or its activity.
(v) The gradient medium should be available as a pure compound and be cheap to use either because it is inexpensive or because it can be recovered after use.

In view of these stringent requirements for an ideal medium, it is not surprising that no one compound is able to meet all of these criteria.

For rate-zonal separations, sucrose meets most of the criteria for an ideal gradient medium. However, in the case of isopycnic separations no one medium has proved satisfactory for all types of biological particles. Hence a wide range of gradient media have been used for different types of biological sample and the applications of a selection of these are shown in *Table 1*. For historical reasons sucrose has been widely used for the isopycnic separation of

Table 1. Applications of the Various Types of Isopycnic Gradient Media.[a]

Media	DNA	RNA	Nucleo-proteins	Mem-branes	Subcellular organelles	Cells	Viruses
Sugars (e.g. sucrose)	–	–	+	+ +	+ +	+	+ +
Polysaccharides (e.g. Ficoll)	–	–	–	+	+	+ + +	+ +
Alkali metal salts (e.g. CsCl)	+ + +	+ +	+	–	–	–	+ +
Colloidal silica (e.g. Percoll)	–	–	–	+	+ +	+ +	+
Nonionic iodina-ted compound (e.g. Nycodenz)	+ +	+ +	+ + +	+ + +	+ + +	+ +	+ +

[a]The classification is as follows: + + +, ideal; + +, satisfactory; +, limited applications; –, un suitable.

membranes and subcellular organelles. In cases where the high osmolarity of sucrose is a problem, polysaccharides (e.g., Ficoll) can be used, although they do form very viscous solutions. Another type of gradient medium developed specifically for characterising nucleic acids were the salts of alkali metals, principally CsCl and Cs_2SO_4 (4) although other salts have also been used (5). The use of such gradients of concentrated salt solutions has been restricted mainly to the separation of macromolecules, although some fractionations of salt-resistant nucleoprotein complexes have also been carried out (5). Another major class of isopycnic gradient media are the colloidal silica suspensions, in which interest has been renewed following the introduction of Percoll. The surface of the silica particles of Percoll are coated to minimise their toxicity and interaction with membranes (6). Even so a number of problems have been encountered in that the presence of ions or low pH causes the colloid to become destabilised and the silica particles sediment rapidly. Also, it can be difficult to remove the silica particles from the samples after fractionation, especially from cells, some types of which appear to ingest the particles of silica (7).

Because of the serious deficiencies of the other classes of gradient media, another class of compounds, the iodinated aromatic compounds, originally devised for X-ray contrast work, have been investigated as possible compounds for isopycnic gradient media. A previous article has dealt in detail with the various types of X-ray contrast media that have been used and their applications (3). This chapter concentrates on the properties of the two nonionic iodinated compounds, metrizamide and Nycodenz, comparing their properties with those of their ionic counterpart, metrizoic acid. The wide range of applications of the nonionic iodinated compounds (*Table 1*) show them to be the most versatile of the various media currently available.

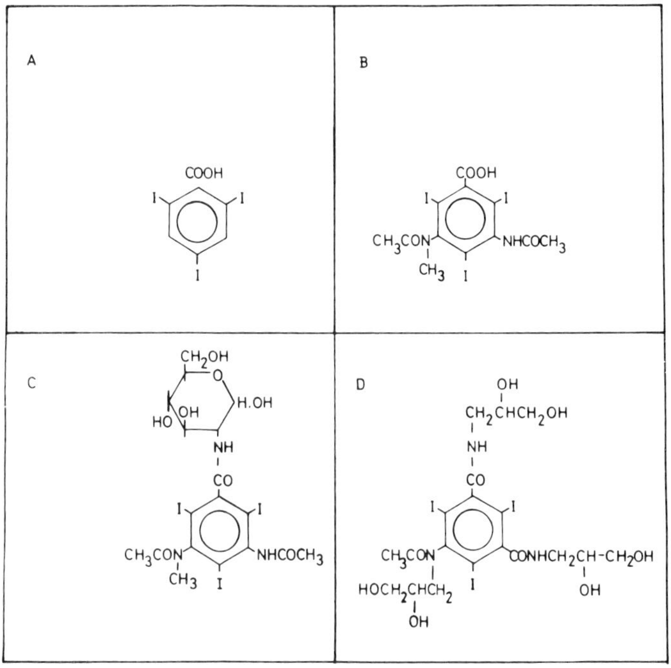

Figure 1. Chemical structures of iodinated compounds. (A) Tri-iodobenzoic acid; (B) metrizoic acid; (C) metrizamide; (D) Nycodenz.

2. CHEMICAL STRUCTURES

Most iodinated compounds used as gradient media have a structure based on tri-iodobenzoic acid (*Figure 1A*) to which hydrophilic groups are attached to increase the solubility of these compounds in water. A detailed description of the various ionic iodinated compounds that have been used has been published previously (3). The structures of most of these various ionic compounds are very similar to that of metrizoic acid (*Figure 1B*) although others (e.g., ioglycamic acid) have a dimeric type of structure. The important difference seen when comparing metrizoic acid with metrizamide (*Figure 1C*) and Nycodenz[1] (*Figure 1D*) is that in the case of the last two compounds the carboxyl group is blocked by covalent linkage to a glucosamide and a dihydroxypropylamine group, respectively. These covalent bonds have a dramatic effect on the properties of these compounds in terms of their stability and suitability as gradient media.

3. PHYSICO-CHEMICAL PROPERTIES OF SOLUTIONS

Salts of metrizoic acid, but not free metrizoic acid itself, are soluble in water; metrizoates are usually obtained as solutions. While the sodium salts are very soluble, the presence of calcium or magnesium ions in the gradient can lead to

[1]Nycodenz is a registered trade mark of Nyegaard & Co., (Oslo, Norway).

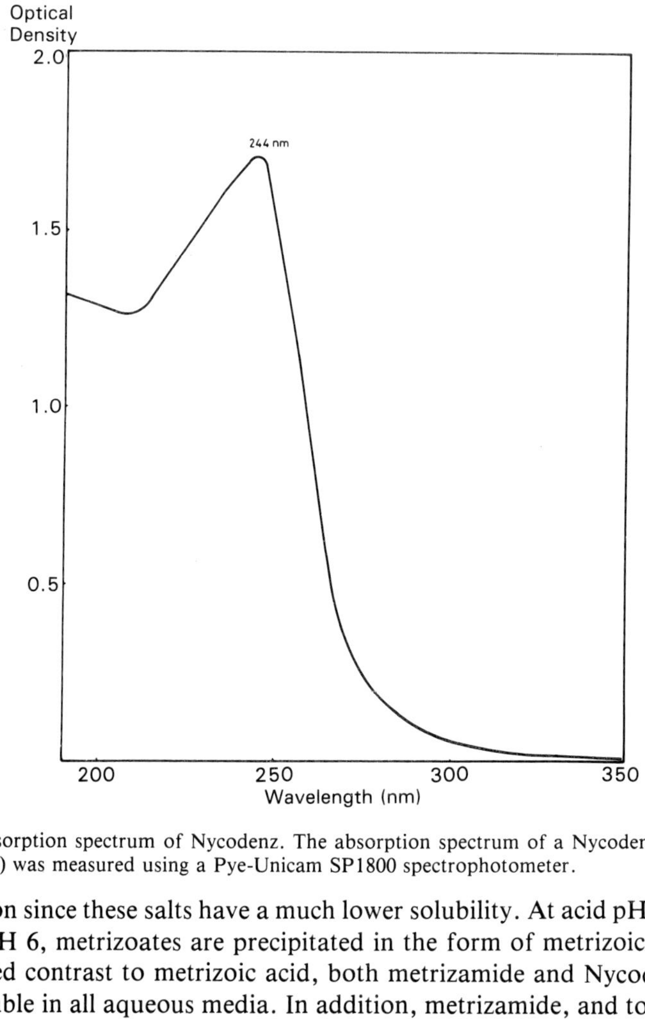

Figure 2. Absorption spectrum of Nycodenz. The absorption spectrum of a Nycodenz solution (5 mg/100 ml) was measured using a Pye-Unicam SP1800 spectrophotometer.

precipitation since these salts have a much lower solubility. At acid pH, that is, less than pH 6, metrizoates are precipitated in the form of metrizoic acid.

In marked contrast to metrizoic acid, both metrizamide and Nycodenz are readily soluble in all aqueous media. In addition, metrizamide, and to a lesser extent Nycodenz, are also soluble in organic solvents such as ethanol, formamide and dimethyl sulphoxide. In marked contrast to the ionic metrizoates, the solubilities of metrizamide and Nycodenz are not affected either by the ionic environment or by the pH of the solution. Studies have shown that both metrizamide and Nycodenz are soluble and stable over the range pH 2 to pH 12.5. Solutions of sodium metrizoate and Nycodenz are stable to heat and they can be sterilised by autoclaving. However, the presence of the glucosamide group in metrizamide makes it unstable to temperatures above 55°C and so metrizamide solutions must be sterilised by filtration. The presence of a tri-iodinated benzene ring in all of these compounds results in a high absorption in the ultraviolet region of the spectrum (*Figure 2*).

These iodinated compounds exhibit a number of favourable properties as gradient media. *Tables 2, 3* and *4* show the physico-chemical properties of solutions of sodium metrizoate, metrizamide and Nycodenz, respectively and

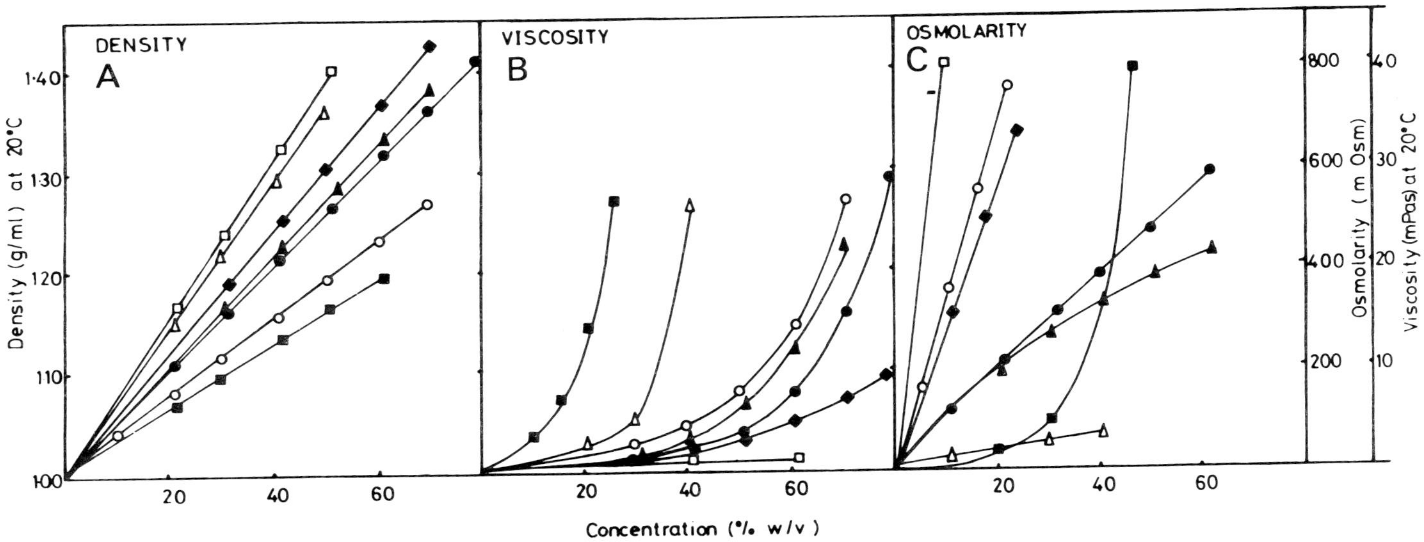

Figure 3. A comparison of physico-chemical properties of gradient media. The relationships between the concentration (% w/v) of gradient solutes and their (A) density (B) viscosity and (C) osmolarity are shown for Nycodenz (●-●), metrizamide (▲-▲), sodium metrizoate (◆-◆), CsCl (□-□), sucrose (○-○), Ficoll (■-■) and Percoll (△-△).

Table 2. Physico-chemical Properties of Sodium Metrizoate Solutions.

Concentration (% w/v)	Concentration (mol/l)	Refractive index (η) 20C°	Density (g/ml) 20°C	Viscosity (mPas)	Osmolarity (mOsm)
0	0.000	1.3330	0.9982	0	1.0
10	0.154	1.3487	1.0601	317	1.3
20	0.307	1.3648	1.1223	631	1.6
30	0.406	1.3809	1.1836	980	2.1
40	0.614	1.3971	1.2449	–	2.8
50	0.768	1.4132	1.3062	–	4.4
60	0.922	1.4293	1.3676	–	6.1
70	1.075	1.4454	1.4289	–	10.1
80	1.229	1.4615	1.4902	–	17.6

Table 3. Physico-chemical Properties of Metrizamide Solutions.

Concentration (% w/v)	Concentration (mol/l)	Refractive index (η) 20°C	Density (g/ml) 20°C	Viscosity (mPas)	Osmolarity (mOsm)
0	0.000	1.3330	0.9982	0	1.0
10	0.127	1.3483	1.0512	107	1.3
20	0.253	1.3646	1.1062	180	1.6
30	0.380	1.3809	1.1612	247	2.3
40	0.507	1.3971	1.2162	320	3.6
50	0.633	1.4133	1.2712	385	6.0
60	0.760	1.4295	1.3262	440	11.0
70	0.887	1.4456	1.3812	–	26.0
80	1.014	1.4620	1.4362	–	58.0

Table 4. Physico-chemical Properties of Nycodenz Solutions.

Concentration (% w/v)	Concentration (mol/l)	Refractive index (η) 20°C	Density (g/ml) 20°C	Viscosity (mPas)	Osmolarity (mOsm)
0	0.000	1.3330	0.999	1.0	0
10	0.122	1.3494	1.052	1.3	112
20	0.244	1.3659	1.105	1.4	211
30	0.365	1.3824	1.159	1.8	299
40	0.487	1.3988	1.212	3.2	388
50	0.609	1.4153	1.265	5.3	485
60	0.731	1.4318	1.319	9.5	595
70	0.853	1.4482	1.372	17.2	1045
80	0.974	1.4647	1.426	30.0	–

Figure 3 shows how the properties of iodinated gradient media compare both with each other and with other types of density-gradient media. For any given concentration, sodium metrizoate solutions are denser and less viscous than equivalent solutions of metrizamide and Nycodenz. However, the osmolarity of sodium metrizoate solutions is much higher than those of the nonionic iodinated media. As can be seen from *Figure 3*, there are no really significant differences between the physico-chemical properties of solutions of metrizamide and Nycodenz.

In comparison with other gradient media, iodinated compounds have a number of advantages. For example, at all densities iodinated gradient media exhibit much lower osmolarities and viscosities than sucrose (*Figure 3*) while polysaccharides form even more viscous solutions. Although CsCl solutions have high densities and low viscosities their use for many isopycnic separations is precluded because of their high osmolarity and, even more serious, their high ionic strength which can dissociate, denature or otherwise disrupt the integrity of biological samples.

4. FORMATION OF GRADIENTS

Gradients of iodinated media can be preformed and the sample loaded onto the top of the gradient. This method is preferred for the preparation of isotonic gradients and also usually for the separation of large particles (e.g., cell organelles) which need to be centrifuged for only a short period of time (i.e., 2 h or less). Alternatively, for small particles, such as macromolecules and nucleoproteins, it is possible to self-form the gradients during centrifugation. In this case the sample is mixed throughout the gradient solution; this not only makes it possible to use a large sample volume but it also removes the possibility of artifactual peaks arising as a result of material remaining in the sample zone.

4.1. **Preformed Gradients**

Gradients can be preformed using any of the standard methods (8). Besides using a simple gradient maker, smooth gradients can be formed by allowing step gradients to diffuse overnight in an upright position or in an hour or less by sealing the top of the tube and allowing the gradient to diffuse while the tube is laid horizontal (*Figure 4*). Another useful, but generally neglected, way of forming gradients is to use the freeze-thaw technique. When a uniform solution of metrizamide or Nycodenz is frozen and then thawed the gradient solute redistributes to give a useful gradient (*Figure 5*); the precise method used to generate the Nycodenz gradients shown in *Figure 5* is given in the figure legend. However, it is not possible to prepare isotonic gradients by the freeze-thaw method; these can only be prepared by using either a gradient maker or the diffusion method. Isotonic gradients of metrizamide and Nycodenz can be prepared with densities up to 1.18 g/ml and 1.15 g/ml, respectively. Although it is possible to use either sucrose or NaCl as the osmotic balancer for isotonic

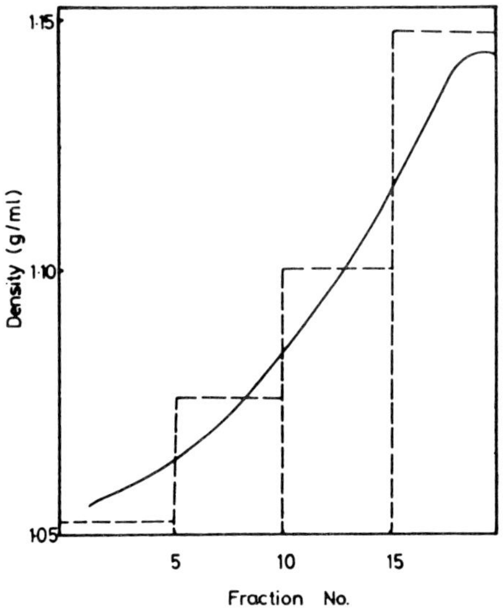

Figure 4. Generation of density gradients by diffusion. A four-step Nycodenz gradient was prepared in a 14 ml (1.0 cm diam.) centrifuge tube. After sealing the top of the tube, the tube was laid on its side for 45 min to allow the gradient to diffuse. The gradient was fractionated and the gradient profile (——) determined from refractive indices. The histogram (– – –) shows the concentration of the original solutions used to prepare the gradient.

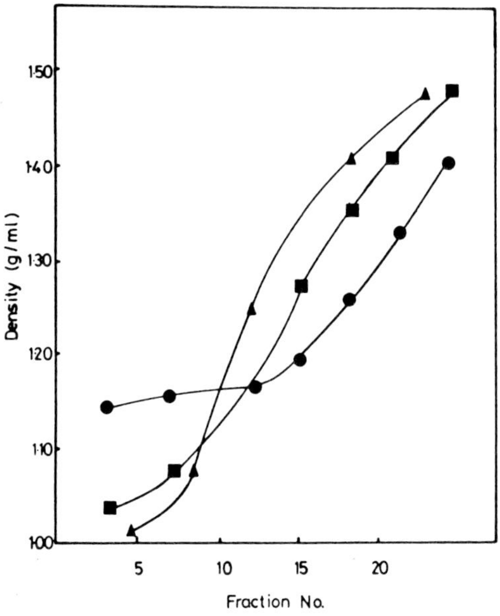

Figure 5. Preparation of gradients using the freeze-thaw technique. Solutions of 5.0 ml of 40% (w/v) Nycodenz in MSE 10 x 10 ml centrifuge tubes were placed in a test-tube rack and frozen at –20°C in a deep-freeze. When frozen the tubes were removed and allowed to thaw out in air at room temperature (approx. 20°C). The solutions were frozen and thawed once (●-●), twice (■-■) or three times (▲-▲).

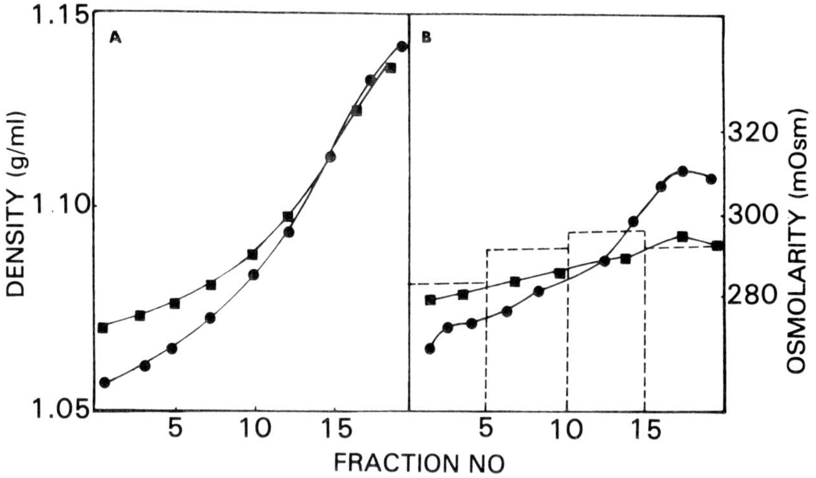

Figure 6. Comparison of the stability of isotonic Nycodenz gradients prepared using either NaCl or sucrose as the osmotic stabiliser. Gradients of 9–27% Nycodenz were layered and allowed to diffuse in the horizontal position for 45 min at room temperature. In each case the density profiles of NaCl-Nycodenz (●-●) and sucrose-Nycodenz (■-■) gradients were measured (A). Similarly the osmotic profiles of NaCl-Nycodenz (●-●) and sucrose-Nycodenz (■-■) gradients were measured (B); the histogram (— — —) shows the osmolarity of the original solutions.

Table 5. Composition of Isotonic Nycodenz Solutions.

Buffered Medium	
	5 mM Tris-HCl, (pH 7.5), containing 3 mM KCl and 0.3 mM $CaNa_2EDTA$ Refractive index (20°C): 1.3331 Osmolarity: 19 ± 1 mOsm
Nycodenz in Isotonic Solution	
	27.6 g solid Nycodenz dissolved in the above medium and made up to 100 ml with that medium Refractive index (20°C): 1.3784 Osmolarity: 290 ± 1 mOsm Density (20°C): 1.15 g/ml
Diluent Solutions	
	0.75 g NaCl or 7.45 g sucrose dissolved in the buffered medium and made up 100 ml with medium
	NaCl diluent (diluent A) Refractive index (20°C): 1.3345 Osmolarity: 250 ± 1 mOsm Density (20°C): 1.003 g/ml
	Sucrose diluent (diluent B) Refractive index (20°C): 1.3446 Osmolarity: 251 ± 1 mOsm Density (20°C): 1.027 g/ml

Table 6. Properties of Nycodenz-NaCl Gradient Solutions.

% (w/v) Nycodenz	27.6	18.4	13.8	9.2
Dilution ratio Nycodenz: Diluent A[a]	1:0	2:1	1:1	1:2
Observed osmolarity (mOsm)	291	295	291	283
Density (g/ml) 20°C	1.146	1.098	1.075	1.050
Refractive index 20°C	1.3784	1.3633	1.3563	1.3490

[a]The composition of diluent A is described in *Table 5*.

Table 7. Properties of Nycodenz-sucrose Gradient Solutions.

% (w/v) Nycodenz	27.6	18.4	13.8	9.2
Dilution ratio Nycodenz: Diluent B[a]	1:0	2:1	1:1	1:2
Observed osmolarity (mOsm)	290	295	290	280
Density (g/ml) 20°C	1.146	1.105	1.086	1.066
Refractive index 20°C	1.3784	1.3669	1.3613	1.3553

[a]The composition of diluent B is described in *Table 5*.

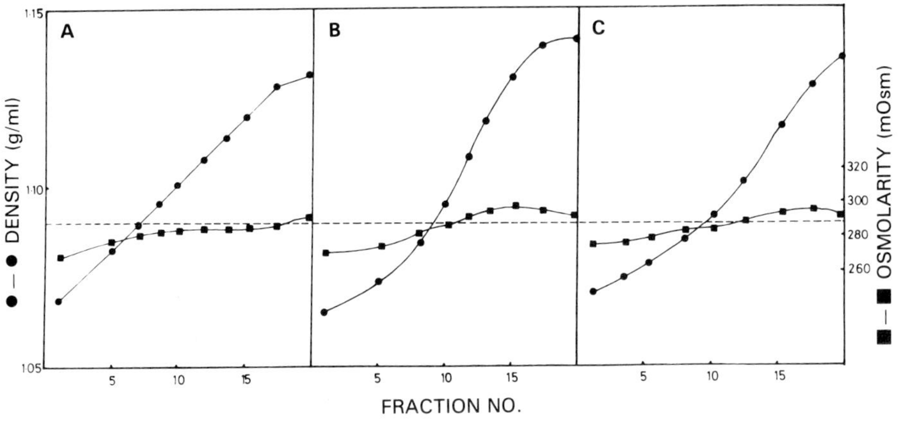

Figure 7. Comparison of isotonic Nycodenz-sucrose gradients produced using (A) a gradient mixer; (B) a two-step diffusion gradient (90 min at 20°C) and (C) a four-step diffusion gradient (45 min at 20°C). The density (●-●) and osmotic (■-■) profiles were measured in each case.

gradients (9, 10) it has been found that, because it diffuses more slowly, sucrose gives more stable gradients (*Figure 6*). Of the iodinated density gradient media, Nycodenz is the most convenient to use because not only are the osmolarities of its solutions lower than those of its ionic counterparts but also, unlike solutions of metrizamide, Nycodenz solutions can be sterilised by autoclaving. The recipes for preparing isotonic gradient solutions of Nycodenz are given in *Table 5; Tables 6* and *7* show the properties of these solutions and *Figure 7* shows the typical gradients and osmotic profiles that can be obtained using either a gradient maker or the diffusion method to prepare Nycodenz-sucrose isotonic gradients. Some types of cells tend to aggregate during centrifugation; this hinders fractionation and gives low yields of cells when the gradient is fractionated. However, this effect can be eliminated completely by the addition of 2% donor calf bovine serum to the gradient solutions. In some cases it is possible to enhance the separation of different cell types by using slightly hypo- or hypertonic gradients and this is discussed further in Chapter 7.

Once formed the osmotic profile of the gradient is stable for several hours as long as the tube is kept in the vertical position. The density profiles of both metrizamide and Nycodenz preformed gradients are even more stable and, even when centrifuged at high speed in a swing-out rotor, the gradient remains stable for at least 24 h (*Figure 8*).

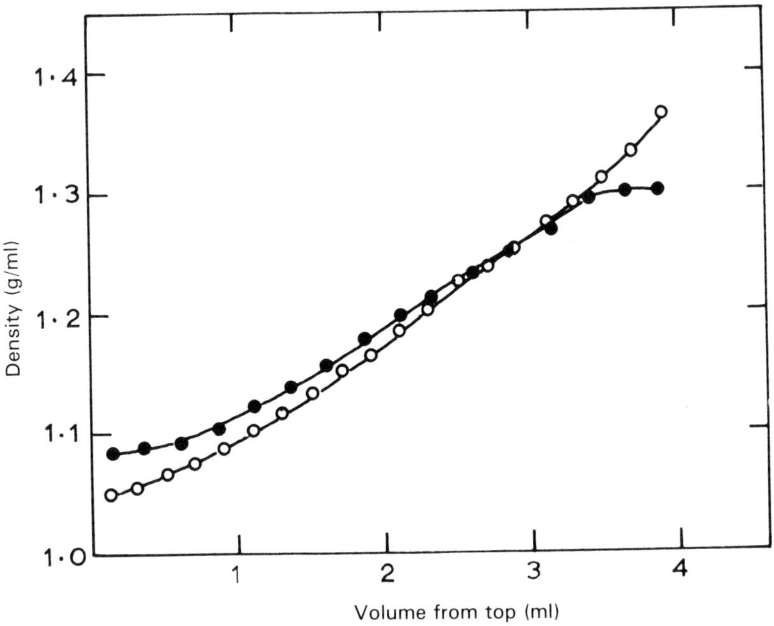

Figure 8. Stability of a preformed metrizamide gradient centrifuged in a swing-out rotor. Gradients were preformed by overlayering 1.0 ml each of 14%, 29%, 43% and 58% (w/v) metrizamide solutions which were then allowed to equilibrate at 4°C for 24 h. The gradients so formed were centrifuged at 100 000xg for 35 h at 4°C in a MSE 3 x 5 ml swing-out rotor. Gradients were fractionated and analysed before (●—●) and after (○—○) centrifugation.

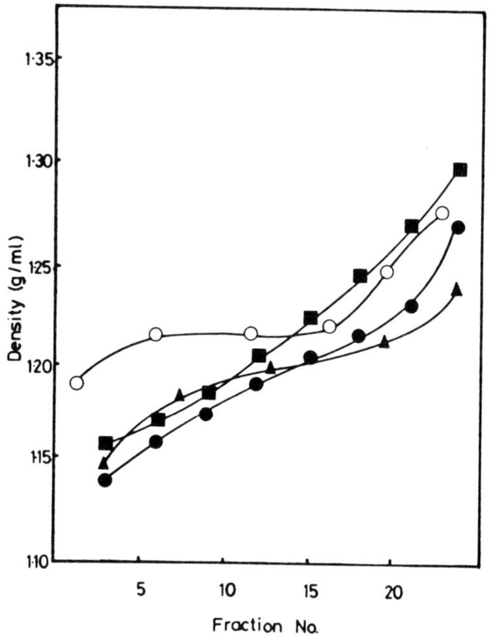

Figure 9. The effect of rotor type on the formation of gradients. Solutions of 40% (w/v) Nycodenz were centrifuged at speeds equivalent, in each case, to 63 000xg, for 24 h at 5°C, in an MSE vertical tube rotor fitted with 5.5 ml adaptors (■-■), an MSE 10 x 10 ml titanium rotor, angle 35° (▲-▲), an MSE 10 x 10 ml aluminium rotor, 21° (●-●) or an MSE 3 x 6.5 ml swing-out rotor (○-○).

4.2. Self-forming Gradients

When solutions of iodinated density-gradient media are centrifuged a gradient begins to form. While a number of parameters affect the rate at which the gradient forms, the most important factor is the sedimentation pathlength. Hence gradients form fastest in vertical rotors in which the pathlength is the diameter of the tube and slowest in swing-out rotors in which the pathlength is the length of the tube; in fixed-angle rotors the rate of gradient formation is intermediate between that of the vertical and swing-out rotors and depends on the angle of the tube in the rotor (*Figure 9*). Although the rate at which gradients form is slower than CsCl gradients, useful gradients can still be generated by centrifuging solutions of Nycodenz in a vertical rotor at 200 000xg for just a few hours (see Appendix to Chapter 6). Other factors apart from the type of rotor and time affect the shape and rate of gradient formation, in order of importance these are the centrifugal force applied, the initial concentration of solution and, in the case of the more viscous gradients of metrizamide and Nycodenz, the temperature during centrifugation (*Figure 10*). By careful choice of rotor and the centrifugation conditions it is possible to manipulate the gradient profile to optimise the resolution of the gradients.

5. ANALYSIS OF GRADIENTS

The analysis of gradients after fractionation consists of two separate processes; it is necessary to measure the density of each fraction and then to determine the distribution of sample throughout the gradient fractions.

5.1. Determination of the Density of Gradient Fractions

There is a linear relationship between the density and refractive index of solutions (*Tables 2–4*). Thus the most convenient method of determining the density of solutions is from their refractive indices because this is quick and easy to measure, it requires only 20 μl or less of each fraction and, when applied correctly, it gives very reliable and accurate results. The equations that have been derived are:

Sodium metrizoate:

$$\text{density (g/ml)} = 3.800\ \eta_{20^\circ} - 4.064$$

Metrizamide:

$$\text{density (g/ml)} = 3.350\ \eta_{20^\circ} - 3.462$$

Nycodenz:

$$\text{density (g/ml)} = 3.242\ \eta_{20^\circ} - 3.323$$

Before applying any of these equations, it is necessary to correct the observed refractive index to take into account the presence of any other solutes (e.g., salts or buffers) since their presence increases the refractive index of the solution. In the case of isotonic Nycodenz gradients, the density of fractions can be estimated using the following equations:

Nycodenz-sucrose:

$$\text{density (g/ml)} = 3.553\ \eta_{20^\circ} - 3.751$$

Nycodenz-NaCl:

$$\text{density (g/ml)} = 3.287\ \eta_{20^\circ} - 3.383$$

However, because these gradients both contain two major gradient solutes each with different distributions in the gradient, different diffusion rates and a different relationship between concentration and refractive index, the density values obtained are less accurate. Even so, the refractive index measurements from fractions of isotonic gradients usually give densities that are accurate enough for most purposes; if a more accurate determination is required then this can be done by pycnometry.

5.2. Analysis of Biological Material in Gradient Fractions

Whenever possible, it is preferable to measure the distribution of material in gradient fractions without removing the gradient medium, since this minimises the possibility of variable losses of material occurring during removal of the gradient medium.

5.2.1. *Spectrophotometry*

One of the most direct methods of analysing a gradient is using spec-

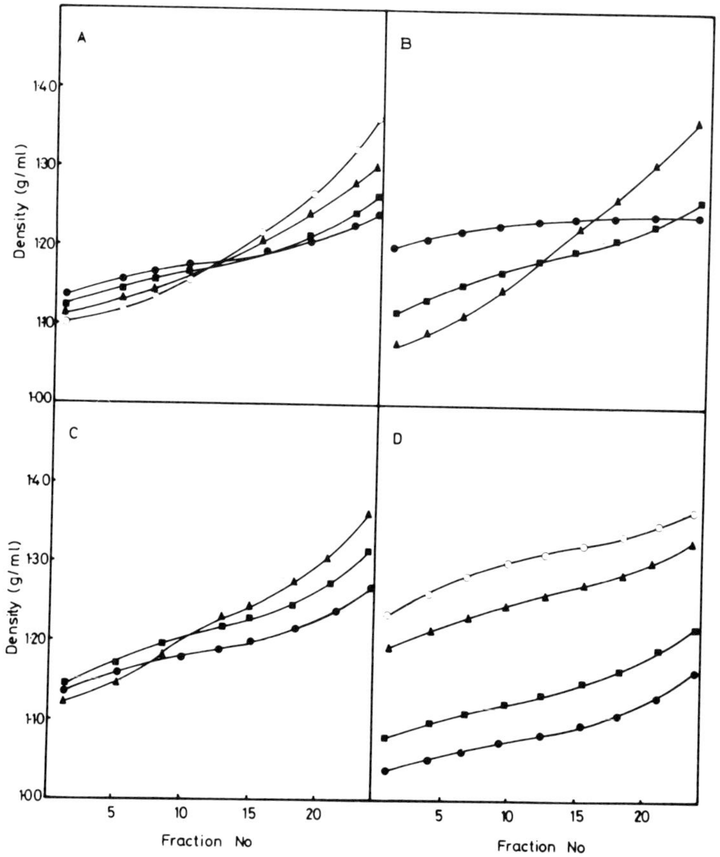

Figure 10. Effect of centrifugation conditions and initial density on the formation of Nycodenz gradients. In all cases 5.0 ml of Nycodenz solution was centrifuged in a MSE 10 x 10 ml aluminium fixed-angle rotor. (A) Effect of centrifugation time on self-forming gradients: solutions of 40% (w/v) Nycodenz were centrifuged for 16 h (●-●), 25 h (■-■), 48 h (▲-▲), and 75 h (○-○) at 60 000xg, 5°C. (B) Effect of centrifugation speed on self-forming gradients: solutions of 40% (w/v) Nycodenz were centrifuged for 24 h at 5°C at 15 000xg (●-●), 60 000xg (■-■), or 120 000xg (▲-▲). (C) Effect of temperature on self-forming gradients: solutions of 40% (w/v) Nycodenz were centrifuged for 24 h at 60 000xg at 5°C (●-●), 15°C (■-■), or 25°C (▲-▲). (D) Effect of initial density on self-forming gradients: solutions of 18% (●-●), 28% (■-■), 49% (▲-▲), and 59% (○-○) were centrifuged for 24 h at 60 000xg at 5°C.

trophotometry. In the case of the iodinated density-gradient media it is possible to carry out spectrophotometric measurements at wavelengths down as far as 400 nm. However, iodinated media, like Percoll and some other media, absorb strongly in the ultraviolet region of the spectrum (see *Figure 2*) and hence it is impossible to carry out any spectrophotometric measurements below 400 nm.

5.2.2. *Chemical Assays*

An alternative method for determining the distribution of biological material

Table 8. Compatibility of Iodinated Gradient Media with Chemical Assays.

Assay for	Method	Type of assay[a]	Compatible with			Range of assay	Reference
			Metrizoate	Metrizamide	Nycodenz		
DNA	Diphenylamine[b]	S	No	No	Yes	20 – 200 μg	17
	Methyl green[b]	S	Yes	Yes	Yes	2 – 200 μg	18
	Ethidium bromide[b]	F	Yes	Yes	Yes	2 – 100 ng	19
	Diaminobenzoic acid	F	No	Yes	Yes	1 – 8 μg	20, 21
	DAPI[b]	F	Yes	Yes	Yes	0.5 – 5 μg	22
	Thiobarbituric acid	F	Yes	–	Yes	0.01 – 10 μg	23
	Adriamycin	S	Yes	Yes	Yes	1 – 20 μg	24
RNA	Orcinol[b]	S	No	No	Yes	10 – 200 μg	17
	Ethidium bromide[b]	F	Yes	Yes	Yes	2 – 100 ng	19
Protein	Folin-phenol	S	No	No	No	20 – 500 μg	25
	Microbiuret	S	No	No	No	100 – 500 μg	26
	Filter techniques[b]	S	No	Yes	Yes	0.1 – 50 μg	27, 28
	Solution dye binding[b]	S	No	Yes	Yes	2 – 50 μg	32
	Fluorescamine[b]	F	Yes	Yes	Yes	1 – 50 μg	29
Polysaccharides (determined as hexoses)	Anthrone	S	No	No	No	10 – 200 μg	30
	Phenol/H_2SO_4[b]	S	No	No	Yes	10 – 200 μg	31

[a]S, spectrophotometric assay; F, fluorimetric assay.
[b]Full experimental method is given in Appendix 1.

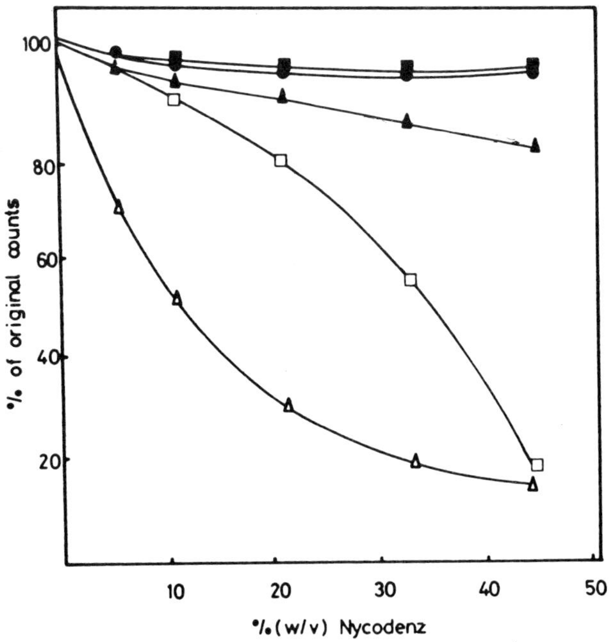

Figure 11. Measurement of radioactivity in the presence of Nycodenz. In each case 100 μl of the appropriate isotope solution was added to 4.5 ml of scintillator and counted for 5 min. Nycodenz solution was then added to each sample to give final concentrations up to 45% (w/v) and the count was repeated. The quenching effect of the different concentrations of Nycodenz was calculated for: ^{14}C (●-●), ^{32}P (■-■) and ^{3}H (▲-▲) in PPO/POPOP/toluene/Triton (13) and for ^{32}P (□-□) and ^{3}H (△-△) in Bray's scintillator (12).

in gradient fractions is to use chemical assays. *Table 8* shows which assays for macromolecules are compatible with sodium metrizoate, metrizamide and Nycodenz. The complete experimental protocols for a selection of assays that are compatible with iodinated gradient media are given in Appendix 1. Sodium metrizoate interferes with any assay that requires acidification of the sample, while the presence of the glucosamine group means that metrizamide interferes with all assays that involve the measurement of sugars (e.g., the orcinol and diphenylamine reactions). Nycodenz is compatible with all of the most important chemical assays and it is possible to measure all types of macromolecules in its presence (11). Nycodenz only interferes with assays which measure amino groups or peptide bonds (e.g., the Lowry and biuret assays). In addition, if the assay conditions are harsh enough (e.g., 98% concentrated H_2SO_4 at 100°C) the molecule decomposes, usually leading to the release of iodine.

5.2.3. *Marker Enzymes*

Marker enzymes are widely used to characterise fractionations of biological material, particularly membrane fractions (11). Most marker enzymes can be assayed in the presence of metrizamide or Nycodenz (*Table 9*), the only proviso is that the assay method should not involve spectrophotometric

Table 9. Effects of Gradient Media on the Activities of Marker Enzymes.

Cell fraction	Marker enzyme[b]	Relative activity[a] in the presence of					
		5% Gradient medium			15% Gradient medium		
		Nyco-denz	Metri-zamide	Sucrose	Nyco-denz	Metri-zamide	Sucrose
Plasma membranes	5′-Nucleotidase	91	108	80	80	100	42
	Adenyl cyclase	100	105	100	64	73	87
	Na^+/K^+-ATPase	48	48	10	55	52	0
Endoplasmic reticulum	NADPH-cytochrome c reductase	82	64	95	50	53	47
	Glucose-6-phosphatase	102	97	nd[c]	84	96	nd
Mitochondria	Succinate dehydrogenase	93	95	108	107	94	121
	Malate dehydrogenase	81	77	nd	36	44	nd
	Mg^{2+}-ATPase	86	91	106	86	93	110
Lysosomes	Acid phosphatase	96	101	105	94	100	nd
Peroxisomes	Catalase	110	105	100	114	118	115

[a]Results are percentages of activity in the absence of any added gradient medium.
[b]Experimental protocols given in Appendix 1.
[c]Not determined.

measurements below 400 nm. Appendix 1 gives the complete experimental protocols for most of the important marker enzymes.

5.2.4. *Radioisotopic Measurements*

Another frequently used method of gradient analysis is radioactive labelling of the sample at some stage prior to fractionation of the sample on a gradient; the distribution of radioactivity across the gradient can then be readily determined by liquid scintillation counting. This technique can be applied to gradients of iodinated density gradient media since they appear to be compatible with all commercial liquid scintillants so far tested (1, 11) in that there is no significant quenching of radioactivity (*Figure 11*). The only scintillant which does give poor counting efficiencies is Bray's scintillant (12) and so this scintillation fluid should be avoided.

In some cases it is desirable to remove the gradient medium before carrying out any further work. There are several possible methods available and these are decribed in the next section.

6. REMOVAL OF IODINATED GRADIENT MEDIA FROM SAMPLES

6.1. **Centrifugation**

Particles down to the size of membrane fractions and subcellular organelles can usually be separated from the gradient medium by centrifugation. To do this, first dilute the gradient fractions with two volumes of an appropriate buffer to reduce the density of the solution below that of the particles. Then centrifuge the diluted gradient fractions; typically membrane fractions need to be centrifuged at 100 000xg for 30 min while larger particles such as cells can be pelleted by centrifugation at 5000xg for 30 min. After centrifugation, wash the pelleted material by resuspension and recentrifugation at least twice in an appropriate buffer to remove the last traces of gradient medium. The main problems usually associated with this type of procedure are that pelleting may cause artifactual changes in the material. For example, particles (e.g., membranes) may aggregate while osmotically sensitive particles (e.g., cells) may be damaged either as a result of the pelleting process or as a result of osmotic stresses during washing of the pelleted material. Fortunately, a number of other methods are available for isolating cells free of gradient medium and these are described in Section 6.4.

6.2. **Dialysis and Ultrafiltration**

The relatively low molecular weights of the iodinated compounds (mol. wts of 600 to 820), means that it is possible to remove them from biological material by dialysis or ultrafiltration. As it has the lowest molecular weight, sodium metrizoate dialyses out faster than metrizamide and Nycodenz. It has been found that there are significant differences in the pore-sizes of dialysis membranes depending on the source of the membranes and hence it is advisable to try several different types to find which type gives best results, especially for use with solutions of metrizamide or Nycodenz. Typically, it takes 24 h or more to dialyse out metrizamide and Nycodenz; somewhat less time is needed to remove sodium metrizoate. An alternative and faster method of removing these gradient media is to use ultrafiltration (11). Using a membrane with a nominal cut-off limit of 5000 daltons. it is possible to remove metrizamide or Nycodenz from samples over the space of a few hours; the actual time depends on the surface area of the membrane and the flow rate used.

6.3. Selective Precipitation and Extraction Procedures

All iodinated gradient media are soluble in solutions of 70% ethanol and 70% acetone and hence it is often feasible to precipitate nucleic acids and proteins selectively using these organic precipitants. To do this, simply add two volumes of ethanol or acetone to the gradient fractions[2] (14). Some time is usually required for a flocculant precipitate to form; if in doubt leave the solution at −20°C overnight. Alternatively, if very small amounts of material are being handled it may be necessary to aid the formation of a precipitate by adding some carrier material which will co-precipitate with the sample but not hamper subsequent analysis. Once the precipitate has formed, pellet it by centrifugation at 10 000xg for 30 min; if only small amounts of material are present then centrifuge at 50 000xg for 30 min. Wash the pelleted material free of any remaining traces of gradient media by resuspension in 90% ethanol or acetone as appropriate and recentrifuge to pellet the precipitate. This washing procedure will need to be done at least once more. The advantage of this method is that it not only removes the gradient medium but it also concentrates the sample. Proteins can also be selectively precipitated by salting out with ammonium sulphate (14). To precipitate all proteins, slowly add 0.7 g of $(NH_4)_2SO_4$ for each millilitre of sample with constant stirring at 5°C. Recover the precipitate by centrifugation at 10 000xg for 30 min then wash the precipitate at least twice with saturated $(NH_4)_2SO_4$ solution by resuspension and recentrifugation. Metrizamide and Nycodenz, but not sodium metrizoate, are also soluble at acid pH and hence some samples (e.g., nucleic acids and proteins) can be acid-precipitated from solution by the addition of trichloroacetic acid to a final concentration of 10% (w/v) or of perchloric acid to a final concentration of 0.6 M.

Nucleic acids separated on gradients are often subsequently purified by extraction with buffered phenol or a phenol-chloroform mixture (15) as part of the experimental protocol. It has been found that phenol extraction is very effective at removing both metrizamide and Nycodenz from gradient fractions. After two or three extractions with phenol essentially all of the gradient medium is extracted from the aqueous phase.

6.4. Miscellaneous Methods

In addition to the general methods for removing iodinated gradient media described in the preceding sections, there are a number of other methods which

[2]In order to optimise the precipitation of nucleic acids sufficient cations must be present; if necessary add a tenth volume of 2.0 M sodium acetate (pH 6.0) prior to addition of the ethanol.

can be used with specific types of biological sample. One example of such a method is the use of affinity chromatography in which case the column matrix has bound to it a particular antibody or group which binds selectively to the sample. Once the sample is bound to the column it can be washed free of gradient medium prior to its elution from the column. In this respect it should be noted that the sugar group of metrizamide can interfere with the interaction between lectins and glycoproteins.

In the case of cells which grow as monolayer cultures, another method of removing metrizamide or Nycodenz is possible (16). The procedure is as follows: put the individual gradient fractions containing cells directly into the wells of a microtitre plate. Add two volumes of tissue culture medium to all of the fractions to ensure that the density of the diluted gradient fractions are less than that of the cells; dilution of the gradient medium to about 10% (w/v) is usually sufficient. Incubate the cells to allow them to establish as a monolayer (approx 24 h), pour off the diluted gradient fractions and add fresh tissue culture medium to the wells. Cells grow in both metrizamide and Nycodenz and most problems arise if the gradient fractions are not diluted sufficiently and the cells are unable to attach to a surface. If this does happen, dilute the gradient fractions with more tissue culture medium and incubate the cells again until they do form a monolayer.

7. RECOVERY OF GRADIENT MEDIA AFTER USE

Although sodium metrizoate is cheap, the cost of metrizamide and Nycodenz gradients tends to be greater; almost as much as CsCl gradients. Hence, some

Table 10. Recovery of Metrizamide and Nycodenz from Gradient Fractions.

1. *Removal of macromolecules*

Separate high molecular weight species using dialysis or ultrafiltration; in the latter case a membrane with a cut-off limit of 5000 daltons. is recommended. Using dialysis it is necessary to use several changes of water over a period of 24-48 h depending on the type of membrane.

2. *Removal of ionic species*

Pass the metrizamide or Nycodenz solution over a small column of mixed-bed resin (e.g. AG 501 (D) from Biorad Laboratories Ltd.) to remove all ionic species.

3. *Concentration*

Concentrate the solution by lyophilisation to dryness. As an alternative it is possible to use a rotary evaporator but in the case of metrizamide the temperature should not exceed 50°C.

4. *Removal of small uncharged molecules*

Dissolve the solid to a final concentration of 30–40% (w/v) and apply it to a Sephadex G-25 column equilibrated with H_2O; the volume of the column should be at least ten times that of the sample. Monitor the column eluate at 244 nm, pool the peak fractions and concentrate as in Step 3.

5. *Verification of purity*

For Nycodenz check the melting point which should be between 174°C and 180°C. For metrizamide there is no simple method; ionic contaminants may be detected by higher than expected conductivities and organic contaminants can be detected by n.m.r. If the final product is unsatisfactory repeat Steps 1–4.

people prefer to recover and reuse these gradient media. The protocol for both types of media is similar and is given in *Table 10*. In the case of metrizamide it is difficult to verify the purity of the recovered material because it decomposes before it reaches its melting point. However, Nycodenz is more stable and the purity of the final product can be verified by checking its melting point which should be between 174° and 180°C.

8. REFERENCES

1. Ridge,D. (1978) in *Centrifugal Separations in Molecular and Cell Biology* (Birnie,G.D. and Rickwood,D. eds.) Butterworths, London and Boston, p. 33.
2. Hames,B.D. (1978) in *Centrifugation: A Practical Approach* (Rickwood,D. ed.) IRL Press Ltd, Oxford and Washington, p. 81.
3. Hinton,R.H. and Mullock,B.M. (1976) in *Biological Separations in Iodinated Density Gradient Media* (Rickwood,D. ed.) IRL Press Ltd, Oxford and Washington, p. 1.
4. Meselson,M., Stahl,F.W. and Vinograd,J. (1957) *Proc. Nat. Acad. Sci. (USA)*, **43**, 581.
5. Birnie,G.D. (1978) in *Centrifugal Separations in Molecular and Cell Biology* (Birnie,G.D. and Rickwood,D. eds.) Butterworths, London and Boston, p. 169.
6. Pertoft,H., Laurent,T.C., Laas,T. and Kagedal,L. (1978) *Anal. Biochem.*, **88**, 271.
7. Wakefield,J.S.J., Gale,J.S., Berridge,M.V., Jordan,T.W. and Ford,H.C. (1982) *Biochem. J.*, **202**, 795.
8. Hames,B.D. (1978) in *Centrifugation: A Practical Approach* (Rickwood,D. ed.) IRL Press Ltd, Oxford and Washington, p. 47.
9. Seglen,P.O. (1976) in *Biological Separations in Iodinated Density Gradient Media* (Rickwood,D. ed.) IRL Press Ltd, Oxford and Washington, p. 107.
10. Ford,T.C. and Rickwood,D. (1982) *Anal. Biochem.*, **124**, 293.
11. Rickwood,D., Ford,T.C. and Graham,J. (1982) *Anal. Biochem.*, **123**, 23.
12. Bray,G.A. (1960) *Anal. Biochem.*, **1**, 279.
13. Rickwood,D. and Klemperer,H.G. (1971) *Biochem. J.*, **123**, 731.
14. Green,A.A. and Hughes,W.L. (1955) *Methods in Enzymol.*, **1**, 67.
15. Kirby,K.S. (1968) *Methods in Enzymol.*, **12B**, 87.
16. Freshney,R.I. (1976) in *Biological Separations in Iodinated Density Gradient Media* (Rickwood,D. ed.) IRL Press Ltd, Oxford and Washington, p. 123.
17. Schneider,W.C. (1957) *Methods in Enzymol.*, **3**, 680.
18. Peters,D.L. and Dahmus,M.E. (1979) *Anal. Biochem.*, **93**, 306.
19. Karsten,U. and Wollenberger,A. (1977) *Anal. Biochem.*, **77**, 464.
20. Setaro,F. and Morely,C.G.D. (1976) *Anal. Biochem.*, **71**, 313.
21. Barth,C.A. and Willershausen,B.S. (1978) *Anal. Biochem.*, **90**, 167.
22. Brunk,C.F., Jones,K.C. and James,T.W. (1979) *Anal. Biochem.*, **92**, 497.
23. Nordling,S. and Aho,S. (1981) *Anal. Biochem.*, **115**, 260.
24. Dalbow,P.G. and Bartuska,B.M. (1979) *Anal. Biochem.*, **95**, 559.
25. Lowry,O.H., Rosebrough,N.J., Farr,A.L. and Randall,R.J. (1951) *J. Biol. Chem.*, **193**, 265.
26. Itzhaki,R.F. and Gill,D.M. (1964) *Anal. Biochem.*, **9**, 401.
27. Schaffner,W. and Weissman,C. (1973) *Anal. Biochem.*, **56**, 502.
28. McKnight,G.S. (1977) *Anal. Biochem.*, **78**, 86.
29. Bohlen,P., Stein,S., Imai,K. and Udenfriend,S. (1974) *Anal. Biochem.*, **58**, 559.
30. Fong,J., Schaffer,F.L. and Kirk,P.K. (1953) *Arch. Biochem. Biophys.*, **45**, 319.
31. Dubois,M., Gilles,K.A., Hamilton,J.K., Rebers,P.E. and Smith,F. (1956) *Anal. Chem.*, **28**, 350.
32. Bradford,M. (1976) *Anal. Biochem.*, **72**, 248.

CHAPTER 2

Analysis of Macromolecules and Macromolecular Interactions Using Isopycnic Centrifugation

T. FORD and D. RICKWOOD

1. INTRODUCTION

Traditionally, macromolecules have been fractionated isopycnically on gradients of concentrated salt solutions which are highly ionic and have a low water activity. Solutions of caesium salts, notably CsCl and Cs_2SO_4, have been used for nucleic acids and, principally, solutions of rhubidium salts (e.g. RbCl) have been used for banding proteins. These solutions can cause partial dehydration and denaturation of some macromolecules and dissociation of macromolecular complexes.

In this chapter the uses of alternative gradient media, based on aromatic iodinated compounds, are described. Macromolecules are highly hydrated in iodinated gradient media because of the high water activity of these solutions (1). Moreover, because Nycodenz and metrizamide are both completely nonionic, it is possible to examine the effects of very low concentrations of ions and molecules on the hydration of these macromolecules as shown by changes in their buoyant densities. In addition it is possible to examine the interaction between macromolecules leading to the formation of stable complexes in a wide range of ionic environments.

2. BANDING OF NUCLEIC ACIDS

2.1. **Introduction**

In iodinated density-gradient media nucleic acids band at low densities. The buoyant density of DNA, range 1.10 – 1.17 g/ml, shows that it is highly hydrated. Calculations show (*Table 1*) that the degree of hydration in metrizamide and Nycodenz gradients is 68 moles of H_2O/mole of nucleotide and 64 moles of H_2O/mole of nucleotide, respectively. Hence nucleic acids are marginally less hydrated in Nycodenz solutions. Denatured DNA and RNA both band at higher densities than native DNA in both metrizamide and Nycodenz gradients (*Table 1*). A similar order of increasing densities of native DNA, denatured DNA and RNA are also observed in CsCl gradients. However, in CsCl gradients nucleic acids band at much higher densities because the water activity of these solutions is so low that there is very little water available for the hydration of macromolecules (*Table 1*). One advantage of CsCl gradients is that the limited amount of water available leads to the differential hydration of DNA's of different base compositions, hence allowing one to separate DNA's on the basis of their base compositions (2); such separations do not appear to be possible in gradients of metriz-

Table 1. Buoyant Densities and Hydration of Nucleic Acids in CsCl and Nonionic Iodinated Gradient Media.

Type of nucleic acid[a]	CsCl		Metrizamide		Nycodenz	
	Density (g/ml)	Hydration[b]	Density (g/ml)	Hydration[b]	Density (g/ml)	Hydration[b]
Native DNA	1.700	7	1.118	68	1.126	64
Denatured DNA	1.715	6	1.139	58	1.169	45
RNA	>1.9	3	1.165	46	1.184	40

[a]Refers to DNA containing 40% G + C.
[b]Hydration expressed in terms of moles of water per mole of nucleotide.

Table 2. Procedure for Banding Nucleic Acids.

1. Treat all solutions to be used with chelating resin to remove heavy metal ions.
2. Ensure that the nucleic acid has its correct counterions. DNA or RNA prepared using hydroxyapatite or gradients of caesium salts should be passed over sodium Dowex to give the sodium salt prior to treating the nucleic acid with chelating resin.
3. Mix the solutions of nucleic acids with solutions of metrizamide or Nycodenz (Nyegaard & Co., Oslo, Norway) to give concentrations of 27% (w/v) and 35% (w/v) for DNA and RNA, respectively in the appropriate concentrations of salts and buffer.
4. Centrifuge the gradient solutions in a fixed-angle rotor at 60 000xg for 40 h at 5°C.
5. Fractionate the gradients by upward displacement with Maxidens (Nyegaard & Co., Oslo).
6. Radioactive nucleic acids can be located by scintillation counting. Non-radioactive nucleic acids can be located by one of the assays given in Appendix 1.

amide or Nycodenz unless base-specific DNA-binding ligands are present. *Table 2* describes the typical procedure for banding nucleic acids in gradients of metrizamide and Nycodenz.

2.2. Physical Factors Affecting the Density of Nucleic Acids

The buoyant density of DNA is affected by factors which affect its hydration. In particular, the pH of the solution affects the buoyant densities of both native and denatured DNA (*Figure 1*). In metrizamide gradients between pH 5 and pH 8 the buoyant density of DNA is fairly constant. As the solution becomes more alkaline the buoyant density decreases rapidly between pH 8.5 and pH 10.5; above pH 10.5 the buoyant densities of native and denatured DNA are essentially the same reflecting the denatured state of the native DNA at this high pH (3). As the pH of the solution decreases from pH 4.5 to pH 2.5 the buoyant density of both native and denatured DNA increases significantly. These changes in buoyant density of DNA with changes in pH appear to be a direct effect of changes in hydration of the DNA as a result of the changes in the distribution of charges associated with the DNA molecules.

A limited amount of work has been done on the effect of temperature on the buoyant density of DNA in metrizamide gradients. The studies done (4) suggest that over the temperature range 1°C to 30°C the density of DNA in metrizamide gradients is fairly constant; this is in contrast to the effect seen in CsCl gradients where the density of DNA increases linearly as the temperature increases. Thc

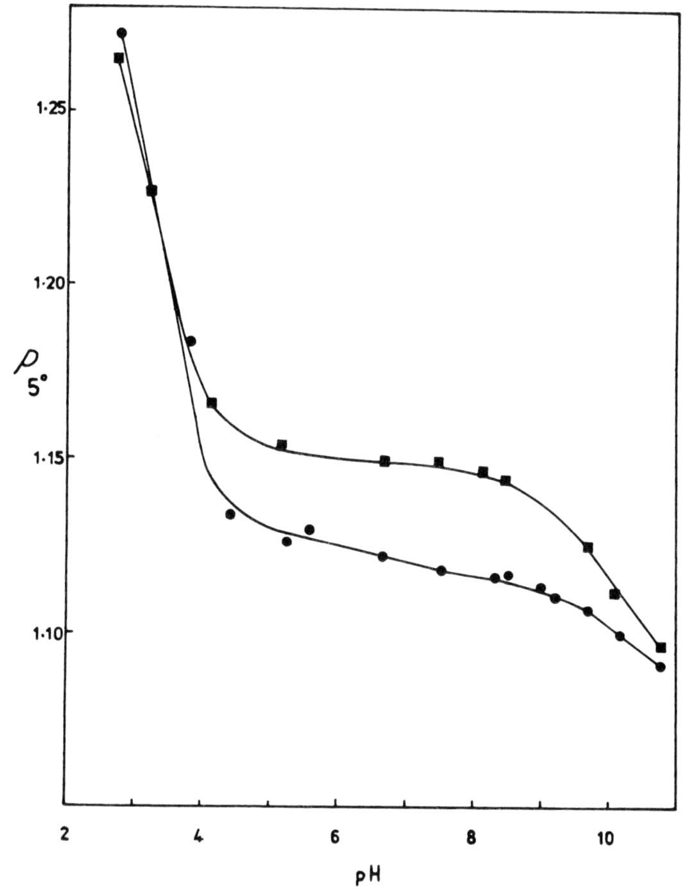

Figure 1. The effect of pH on the buoyant density of DNA. Purified ^{3}H-labelled native (● —— ●) and denatured (■ —— ■) mouse DNA (13) were mixed separately with metrizamide solution to give a gradient solution containing 27% (w/v) metrizamide, 20 mM NaCl, 1 mM EDTA and 10 mM Hepes-NaOH (pH 7.0), the pH of each gradient was then adjusted immediately before centrifugation to give the required pH either by the addition of dilute chelexed NaOH solution or of 'Aristar' HCl. The gradients were centrifuged at 45 000xg for 45 h in an MSE 10 x 10 ml fixed-angle (21°) rotor at 2°C. The gradients were fractionated by upward displacement with Maxidens and the distribution of ^{3}H-DNA was determined by liquid scintillation counting. Data from ref. 4.

reason for this difference remains unclear.

The results described here all relate to metrizamide gradients. However, the great similarity between the behaviour of nucleic acids in metrizamide and Nycodenz gradients make it likely that similar effects will be seen in Nycodenz gradients also.

2.3. Effects of Ionic Environments on the Buoyant Density of Nucleic Acids

2.3.1. *The Ionic Concentration of Gradients*

Increasing the ionic concentration of metrizamide or Nycodenz gradients leads to

Table 3. The Effect of Water Activity on the Buoyant Densities and Hydration of DNA and RNA.

Concentration of NaCl (mM)	a_w^a	Native DNA		Denatured DNA		RNA	
		Buoyant density (g/ml)	Moles H_2O per mole nucleotide	Buoyant density (g/ml)	Moles H_2O per mole nucleotide	Buoyant density (g/ml)	Moles H_2O per mole nucleotide
20	0.9990	1.118	68	1.139	58	1.165	46
100	0.9968	1.123	66	1.142	56	1.206	30
200	0.9932	1.128	63	1.151	52	1.234	25

[a]Calculated from freezing point depression data (20).

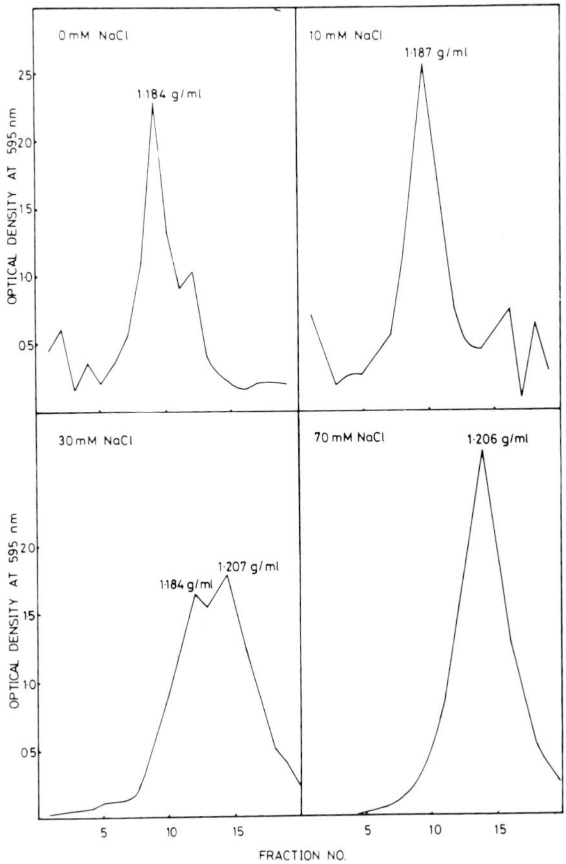

Figure 2. The effect of NaCl concentration on the buoyant density of RNA. Cytoplasmic RNA from *S. cerevisiae* was centrifuged on 35% (w/v) self-forming Nycodenz gradients (5 ml/gradient), containing 10 mM Tris-HCl (pH 7.0) 1 mM EDTA and NaCl concentrations varying between 0–100 mM. Gradients were centrifuged at 63 000xg for 42 h at 5°C in an MSE 10 x 10 ml aluminium rotor. After fractionation the distribution of RNA was determined using the orcinol assay (Appendix 1). The figure shows results with concentrations of NaCl up to 70 mM, increasing the concentration to 100 mM NaCl produced no further change in the buoyant density.

an increase in the buoyant density of both native and denatured DNA as well as of RNA (*Table 3*) reflecting the decreasing amounts of water available for the hydration of nucleic acids as the ionic concentration increases. In the case of RNA the changes in hydration appear to reflect not only the water activity of the solution but also changes in conformation of the RNA in solution. As shown in *Table 3*, as compared to DNA, there is a larger decrease in the hydration of RNA over the range 20 mM NaCl to 100 mM NaCl. This effect has been investigated further using Nycodenz gradients (*Figure 2*). At low ionic strength the RNA bands as a narrow discrete peak at 1.184 g/ml. Increasing the salt concentration to 30 mM NaCl results in some of the RNA banding denser at 1.207 g/ml. At 70 mM NaCl and up to 100 mM NaCl RNA again forms a single, narrow band at 1.206 g/ml (24).

2.3.2. *Effects of Cations and Anions*

As might be expected, cations interact with DNA and they can change both its partial specific volume and its hydration. These changes are reflected in the buoyant density of DNA in metrizamide gradients (*Table 4*), in addition the hydration of DNA, as reflected by its buoyant density, is affected by the anions present in the solution (*Table 5*). In this case it is not clear whether the changes in hydration of the DNA seen, from 60 moles of H_2O/mole of nucleotide to

Table 4. The Effect of Caesium and Sodium Ions on the Buoyant Density and Hydration of Native DNA in Metrizamide Gradients Containing 20 mM Salt.

Salt present	$\bar{V}^a$ (ml/g)	Buoyant density (g/ml)	Hydration (moles H_2O/ mole nucleotide)
NaCl	0.500	1.118	68
CsCl	0.440	1.179	49
Na_2SO_4	0.500	1.133	60
Cs_2SO_4	0.440	1.200	43

[a]From Cohen and Eisenberg (21).

Table 5. The Effect of Anions on the Buoyant Densities and Hydration of DNA and RNA in Metrizamide Gradients Containing 20 mm Salt.

Salt present	Native DNA		Denatured DNA		RNA	
	Buoyant density (g/ml)	Moles H_2O per mole nucleotide	Buoyant density (g/ml)	Moles H_2O per mole nucleotide	Buoyant density (g/ml)	Moles H_2O per mole nucleotide
Na_2So_4	1.133	60	1.159	49	1.212	29
NaF	1.126	64	1.158	49	–	–
NaCl	1.118	68	1.137	58	1.165	46
NaBr	1.109	75	1.130	62	–	–
NaI	1.100	83	1.114	71	1.124	56

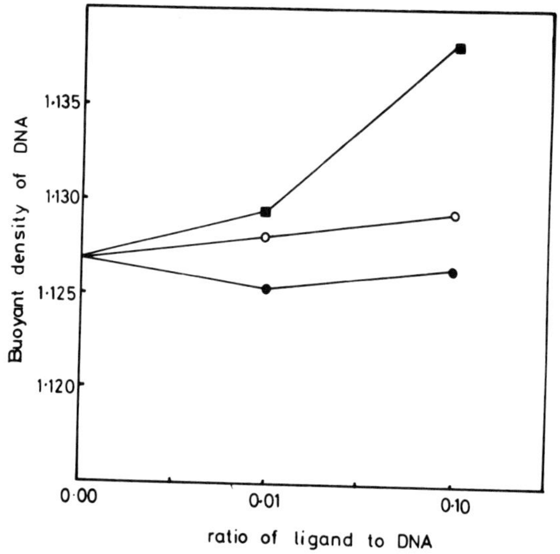

Figure 3. Effect of ligand binding on the buoyant density of DNA in Nycodenz gradients. Calf thymus DNA (200 μg) was banded in gradients of 27% (w/v) Nycodenz containing 1 mM EDTA, 10 mM NaCl, 10 mM Tris-HCl (pH 7.5) and varying amounts of DAPI (■——■), acridine orange (○——○) and ethidium bromide (●——●). The gradients were centrifuged at 63 000xg for 44 h at 5°C in an MSE 10 x 10 ml fixed-angle rotor. The ratio of ligand to DNA is expressed in terms of moles of ligand per base pair.

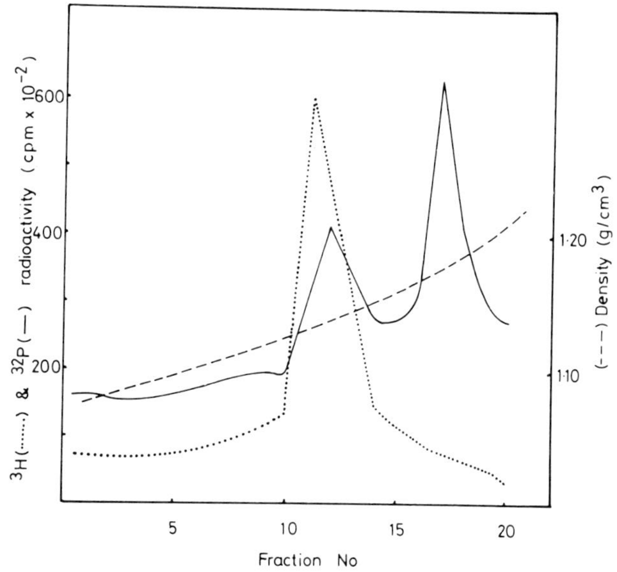

Figure 4. The buoyant densities of native and density-labelled *E. coli* DNA in Nycodenz gradients. *E. coli* ^{3}H-labelled native DNA and ^{32}P-labelled DNA, density-labelled with 5-bromodeoxyuridine, were centrifuged on 27% (w/v) self-forming Nycodenz gradients, containing 10 mM Tris-HCl (pH 7.5) and 1 mM EDTA, at 63 000xg for 44 h at 5°C in an MSE 10 x 10 ml aluminium rotor. After fractionation, the density (— — —) of each fraction was measured and the distributions of native DNA (· · · · ·) and density-labelled DNA (———) were determined by isotopic measurements (24).

83 moles of H_2O/mole of nucleotide, are the result of varying interactions of the anions with the DNA or the result of changes in the water activity of the solutions.

2.4. Effect of Ligand-binding on the Density of DNA

Since DNA bands at a low density in metrizamide and Nycodenz gradients (1,24) one might expect that ligands would not have a large effect on the density of DNA unless they either have a marked effect on the partial specific volume (as in the case of caesium ions) or affect the hydration of the DNA. In fact this is the case and ligands such as acridine orange and ethidium bromide only marginally affect the buoyant density of DNA (*Figure 3*). In contrast 4′,6-diamidino-2-phenylindole (DAPI) increases the buoyant density of DNA (*Figure 3*) presumably because its binding reduces the level of hydration; similarly the binding of sibiromycin to DNA also results in a marginal increase of density in metrizamide gradients (25). In addition it has been found that larger shifts in density, sufficient to fractionate DNA on the basis of base composition, can be obtained using peptides such as netropsin that bind preferentially to regions of DNA rich in A-T base pairs (28).

2.5. Density-labelling Studies of Nucleic Acids

When cells are grown in the presence of bromodeoxyuridine it is incorporated into the DNA in place of thymidine. In CsCl gradients such density-labelled DNA is easily separated from non-density-labelled DNA since it bands at a much higher density. In Nycodenz gradients also bromodeoxyuridine-labelled DNA bands much more densely (*Figure 4*). Other studies have shown that RNA density labelled by the incorporation of mercuriated nucleotides can be separated from non-density labelled RNA by centrifugation in metrizamide gradients (5). This method is extremely useful for isolating newly synthesized nucleic acids from other nucleic acids.

3. BANDING OF PROTEINS

3.1. Introduction

Proteins are much less highly hydrated than nucleic acids and they band at similar densities in ionic media such as CsCl and in nonionic iodinated media. However, proteins do not form single discrete bands in metrizamide gradients; instead a significant amount of the protein bands at a higher density (*Figure 5*). This effect is due to a weak reversible interaction between metrizamide and proteins (6,7). This effect has been observed in the case of all proteins so far studied. A similar, but significantly smaller effect, has been observed when proteins are banded in Nycodenz gradients (*Figure 5*). This suggests that Nycodenz reacts much less with proteins than does metrizamide.

3.2. Effects of Nonionic Gradient Media on Enzyme Activity

All low molecular weight gradient solutes inhibit enzymes in a non-specific manner (8) although the exact mechanism of such inhibition remains unclear.

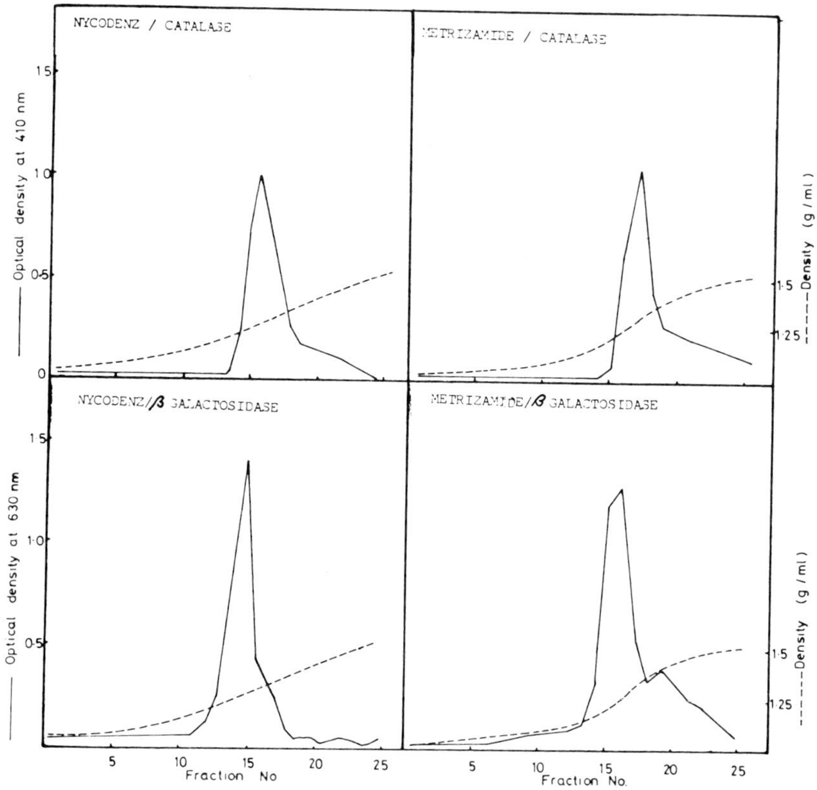

Figure 5. The buoyant density of proteins and their interaction with the gradient media. Protein (200 μg/5 ml gradient) was top-loaded onto self-forming gradients of metrizamide or Nycodenz. Each 5 ml gradient consisted of 4.0 ml 44% (w/v) underlayered with 1.0 ml of 62.5% (w/v) of the gradient medium in 10 mM Tris-HCl (pH 7.5). The gradients were centrifuged in an MSE 10 x 10 ml aluminium rotor at 142 000xg for 62 h at 5°C. After fractionation the density (— — —) of each fraction was determined from its refractive index. The distributions of catalase and β-galactosidase (——) were determined by absorption at 405 nm and the Amido-black assay (Appendix 1), respectively. Data from ref. 24.

Metrizamide and Nycodenz affect the activity of most enzymes to about the same extent as sucrose (see *Table 9* of Chapter 1). However, it has been reported that the glucosamine structure of metrizamide is responsible for the inhibition of hexokinase by metrizamide (9,10). There has not been any other report of specific inhibition of any other enzyme by either metrizamide or Nycodenz.

3.3. Fractionation of Density-labelled Proteins on Metrizamide Gradients

Density-labelled proteins have been fractionated on D_2O/metrizamide gradients (11). The advantage of using D_2O as the gradient solvent is that its greater density (1.105 g/ml) allows one to use lower concentrations of metrizamide. This means that there is less interaction between metrizamide and the proteins, and, in addition, the viscosity of the gradients is less, so proteins reach their isopycnic positions faster.

Table 6. Analysis of Density-labelled Proteins Using D_2O/metrizamide Gradients.

1. *Preparation of sample and gradient solutions.*
   The enzymes or protein mixtures can be applied to the gradient dissolved in H_2O. Dissolve metrizamide powder in D_2O to prepare a stock solution of at least 60% (w/v). Add 2-mercapto-ethanol and Tris-acetate (pH 7.0) to final concentrations of 10 mM and 100 mM, respectively.
2. *Preparation of gradients.*
   Dilute the stock solution with D_2O containing 10 mM 2-mercaptoethanol and 100 mM Tris-acetate (pH 7.0) to give solutions containing 50%, 40%, 30%, and 20% (w/v) metrizamide. Overlay equal volumes of all four gradient solutions (Chapter 1, Section 4.1); the gradient diffuses to linearity during centrifugation. Load the sample onto the gradient in a tenth of the gradient volume (e.g. 0.5 ml for a 5 ml gradient).
3. *Centrifugation conditions.*
   Centrifuge the gradients in a swing-out rotor at the highest possible centrifugal force to obtain best results; typically one uses 350 000xg for 17 h at 4°C. Rotors such as the Beckman SW 65 or MSE 6 x 4.2 ml should give the best results.
4. *Analysis of gradients.*
   For best results fractionate the gradients into 60–70 fractions either by upward displacement with Maxidens or by dripping them out of the bottom of the centrifuge tube. Measure the refractive index and calculate the density using the formula:

$$\text{Density (g/ml)} = 3.0534\,\eta - 2.9541$$

The activity of most enzymes can be determined directly in each fraction without removing the metrizamide. The distribution of non-enzymic proteins can be detected immunologically.

The protocol used for separating density-labelled proteins is given in *Table 6*. This density-labelling technique has been used to study the *de novo* synthesis of enzymes in Myxobacter. The advantage of this method is that, when studying the synthesis of enzymes, complex mixtures of proteins can be loaded onto D_2O/metrizamide gradients and the distribution of the enzyme assayed without prior purification. An example of the type of fractionation that can be achieved is shown in *Figure 6*.

4. BANDING OF POLYSACCHARIDES AND PROTEOGLYCANS

Glycogen bands at fairly high densities in both metrizamide and Nycodenz gradients, 1.28 g/ml and 1.29 g/ml, respectively. In the case of Blue Dextran (Pharmacia Fine Chemicals A/B), in which aromatic chromophores are covalently bound to an uncharged dextran molecule, the presence of the aromatic groups reduces the buoyant density of the molecule resulting in a buoyant density of 1.19 g/ml in both metrizamide and Nycodenz gradients. Highly charged polysaccharides such as hyaluronic acid and chondroitin sulphate appear to be highly hydrated and band at low densities in metrizamide gradients (*Figure 7*); preliminary experiments suggest that these polysaccharides also band at similar densities in Nycodenz gradients. In contrast to hyaluronic acid and chrondroitin sulphate, proteoglycans band at a high density close to that of glycogen (26,27), presumably as a result of the peptides associated with these molecules. The ability of nonionic iodinated gradient media to separate these different types of polysaccharides and proteoglycans has led to the use of metrizamide gradients for study-

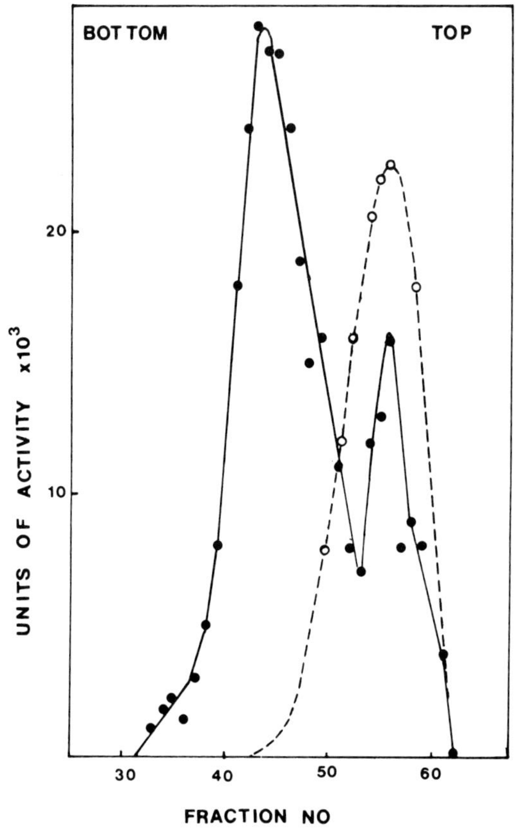

Figure 6. Fractionation of density-labelled proteins in D_2O/metrizamide gradients. Cultures of Myxobacter AL-1 were density labelled by growth in D_2O, extracts from density-labelled cells were separated on metrizamide gradients by centrifugation at 350 000xg for 17 h and the distribution of α-glucosidase (———) was analysed. The banding position of α-glucosidase from unlabelled cultures (— — —) is also shown. Data derived from ref. 11.

ing the metabolism of these compounds (26,27).

Another interesting use of metrizamide gradients has been their use to isolate the complexes of glycogen and its associated enzymes that occur *in vivo* (12). The success of this technique suggests that such gradients may be very useful for the isolation of such macromolecular complexes.

5. INTERACTIONS OF MACROMOLECULES

5.1. **Introduction**

One important area of interest is the interaction between macromolecules, especially interactions between nucleic acids and proteins. Interactions between nucleic acids and proteins can be divided into two main types, namely ionic interactions and sequence specific interactions. Ionic interactions involve the formation of ionic links between basic proteins and the negatively charged phosphate groups of the nucleic acid; these complexes once formed tend to be fairly stable.

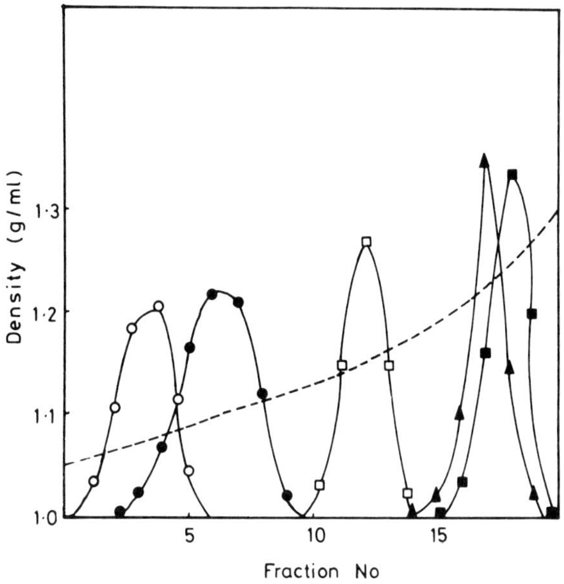

Figure 7. Banding of polysaccharides and proteoglycans in metrizamide gradients. Metrizamide gradients containing chondroitin sulphate (○——○), hyaluronic acid (●——●), porcine proteoglycan (▲——▲), glycogen (■——■) and Blue Dextran (□——□) were centrifuged at 80 000xg for 48 h. Data derived from refs. 24, 26, 27.

The sequence specific interactions between nucleic acids and proteins tend to be less stable, for example, the half-lives of complexes between RNA polymerases and DNA can be in the order of an hour or less, but in other cases completely stable complexes are formed.

The interactions between nucleic acids and proteins have been studied by a variety of methods, especially filter binding techniques and affinity column techniques. The advent of the nonionic iodinated gradient media now allows one to study the formation of stable complexes using a wide range of ionic environments (13,14).

5.2. Interactions Between DNA and Basic Proteins

5.2.1. *Histones*

Histones bind to DNA to form stable complexes. When the mixtures of histones and DNA are dialysed from 2 M NaCl, 5 M urea into 0.14 M NaCl most of the histone molecules become associated with the DNA. The histones appear to bind in a co-operative manner forming at least three types of complexes (13). At low input ratios of histone to DNA two types of complexes containing either half as much histone as DNA (buoyant density: 1.15 g/ml) or equal amounts of histone and DNA (buoyant density: 1.19 g/ml) are observed (*Figure 8*). When the amount of histones present is significantly greater than that of the DNA then a single dense complex with a buoyant density up to 1.25 g/ml and containing up to twice as much histone as DNA is formed (*Figure 8*). All three types of complex contain all

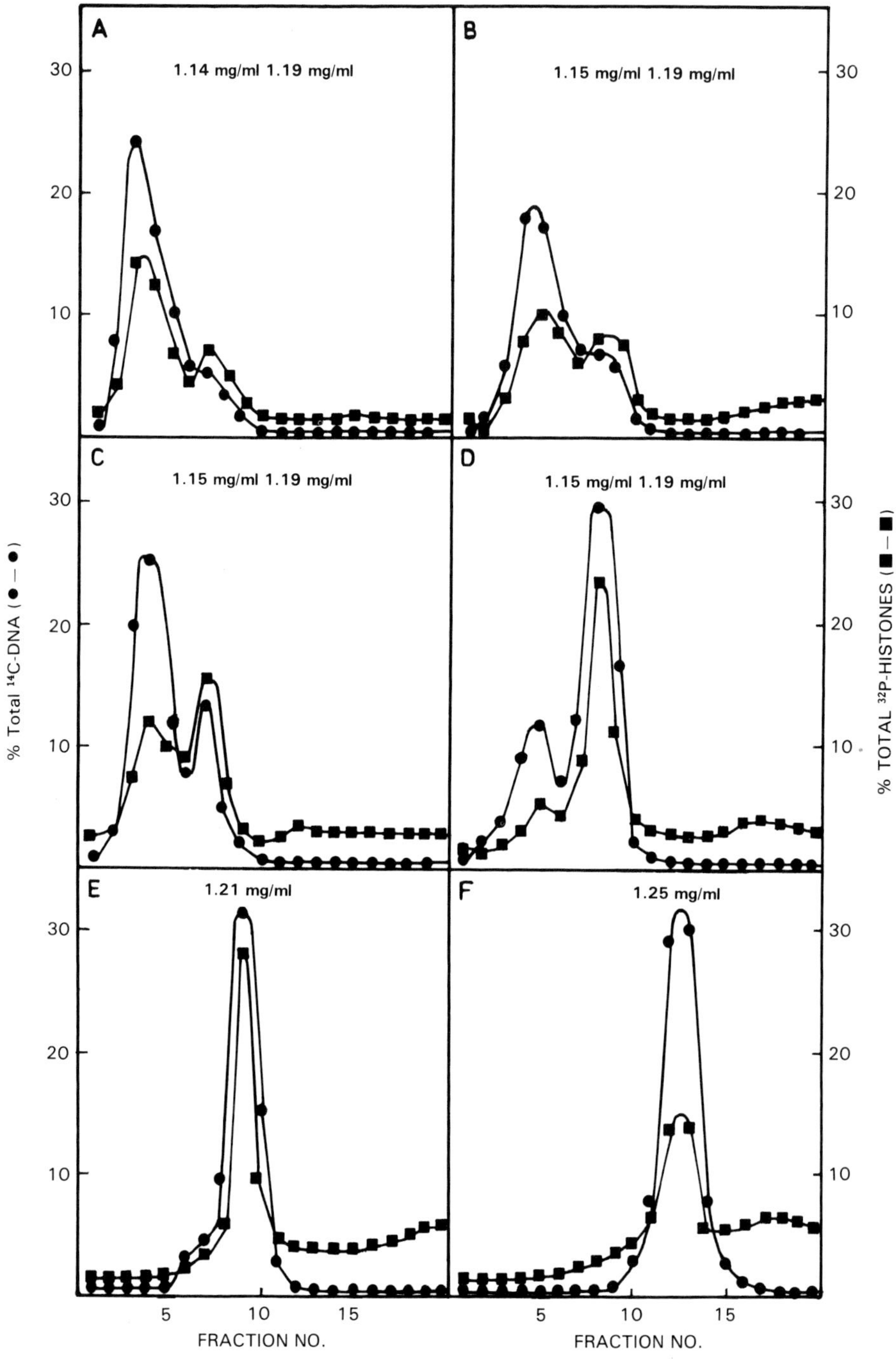

Figure 8. Analysis of the binding of histones to DNA using metrizamide gradients. Mouse DNA (75 μg) and varying amounts of total mouse liver histones were mixed in 2 M NaCl, 5 M urea and reconstituted (13). The complexes formed in 0.14 M NaCl were centrifuged in 35% (w/v) metrizamide gradients at 63 000xg for 44 h at 5°C. The ratios of histone to DNA were (A) 0.2; (B) 0.5; (C) 0.8; (D) 1.0; (E) 1.5; (F) 2.5.

Table 7. Experimental Protocol for Studying the Interaction of Basic Proteins with DNA.

1. *Preparation of polypeptides and DNA*
   Histones should be extracted and purified using one of the more gentle of the standard methods (17, 22), other types of basic protein will require other methods of isolation. The DNA concentration should be carefully chosen to ensure that complexes will band significantly denser than free DNA. Both components are mixed in the required ratio in a solution containing 2 M NaCl, 5 M urea, 1 mM DTT, 0.1 mM EDTA and 10 mM Tris-HCl (pH 7.5).
2. *Dialysis for the binding of proteins to DNA*[a]
   Dialyse the mixture of DNA and proteins at 4°C sequentially against 1.0 M, 0.5 M, 0.25 M and 0.14 M NaCl solutions, also containing 5 M urea, 1 mM DTT, 0.1 mM EDTA and 10 mM Tris-HCl (pH 7.5). Finally remove the urea by dialysis against 0.14 M NaCl, 1 mM DTT, 0.1 mM EDTA and 10 mM Tris-HCl (pH 7.5). The total dialysis time should be 24 h.
3. *Preparation and centrifugation of gradients*
   Mix the samples with an equal volume of a solution containing 70% (w/v) metrizamide (Nyegaard & Co., Oslo, Norway), 0.14 M NaCl, 1 mM DTT, 0.1 mM EDTA and 10 mM Tris-HCl (pH 7.5). Place a 5 ml aliquot of the gradient solution into a 10 ml tube and centrifuge the gradients in an MSE 10 x 10 ml or equivalent fixed-angle (21°) rotor at 63 000xg for 40 h at 2°C.
4. *Analysis of gradients*
   Fractionate the gradients into 20 – 30 fractions, measure the density of each fraction from its refractive index and locate the distribution of DNA and protein either by chemical estimation (see Appendix 1) or, if they are radioactive, by liquid scintillation counting.

[a]If it is wished to study the formation of nucleosomes it is necessary to carry out the reconstitution in the absence of urea.

the major types of histones and similar results are obtained whether eukaryotic or prokaryotic DNA is used (13). It is important to realise that urea was present during the reconstitution of these complexes and that these conditions do inhibit the formation of nucleosomes (15). In fact electron microscopic studies show that the lightest complexes have a fibrillar structure while the denser ones contain spherical aggregates 0.1 – 0.3 μm in diameter. In addition, the results obtained refer to dialysis into 0.14 M NaCl; there is evidence (16) that in the absence of salt discrete complexes are not found. Additional work has shown that reconstitution of purified individual histones to DNA results in the formation of various complexes, the nature of which are dependent on which purified histone fractions are used in the reconstitution procedure (29).

An example of the type of experimental protocol that can be used is given in *Table 7*. It should be noted that the buoyant densities of complexes of DNA and proteins, but not of RNA and proteins, are dependent on the amount of material loaded; this effect is significant in metrizamide gradients but much less so in Nycodenz gradients (see Chapter 4, section 5.1).

5.2.2. *Polylysine and Protamines*

The histones, although a fairly simple mixture of proteins, do show heterogeneity in that some are 'arginine rich' and others are 'lysine rich'. In an attempt to investigate the influence of protein basicity on the formation of protein/DNA complexes the pattern of the binding of polylysine and protamines (75% arginine) has been compared. The method used is the same as that given in *Table 7*. It was

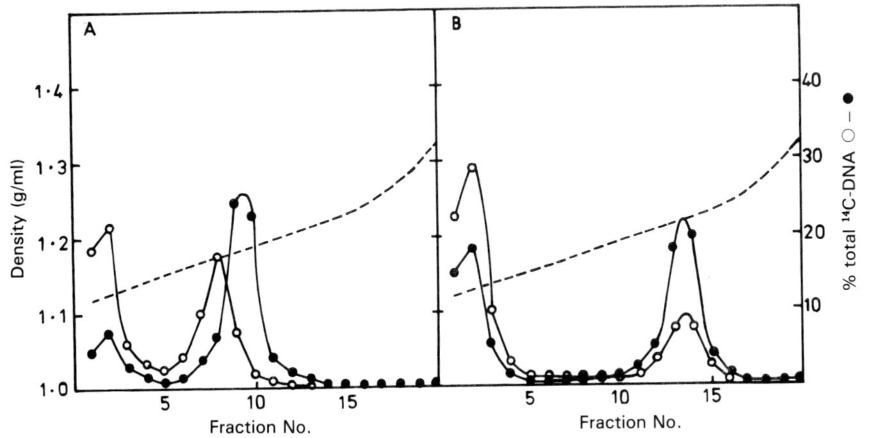

Figure 9. Binding of poly-L-lysine and protamine to mouse DNA. Mouse DNA (75 μg/ml) was reconstituted to poly-L-lysine (A) input ratio of 1.0:0.3 (○——○) and 1.0:0.6 (●——●) and to protamine (B) input ratio 1.0:0.3 (○——○) and 1.0:0.8 (●——●). The reconstituted mixtures were fractionated on metrizamide gradients and the distribution of DNA determined as described in ref. 13.

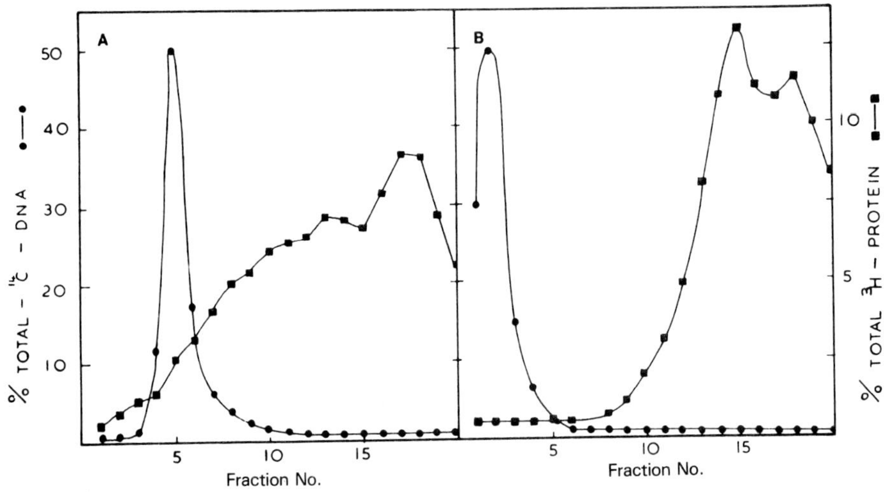

Figure 10. Separation of nucleic acids and proteins on metrizamide gradients. Samples of ^{3}H-labelled proteins (■——■) and ^{14}C-labelled DNA (●——●) were either (A) dispersed throughout the 5 ml gradient and centrifuged separately at 80 000xg for 44 h or (B) loaded into the bottom 1 ml of the gradient and centrifuged separately at 60 000xg for 44 h. Gradients were analysed as described in ref. 4.

found that polylysine binds to and forms complexes with DNA in a similar fashion to total histones; co-operative binding occurs and the density of the complexes increases as the relative ratio of polylysine to DNA increases (*Figure 9*). In contrast, the very basic protamines form dense complexes with DNA even at low ratios of protamines to DNA (*Figure 9*) suggesting that the basicity of the polypeptide, at least in part, determines the nature of the complex formed. Similar effects of basicity are seen when individual histone fractions bind to DNA (29).

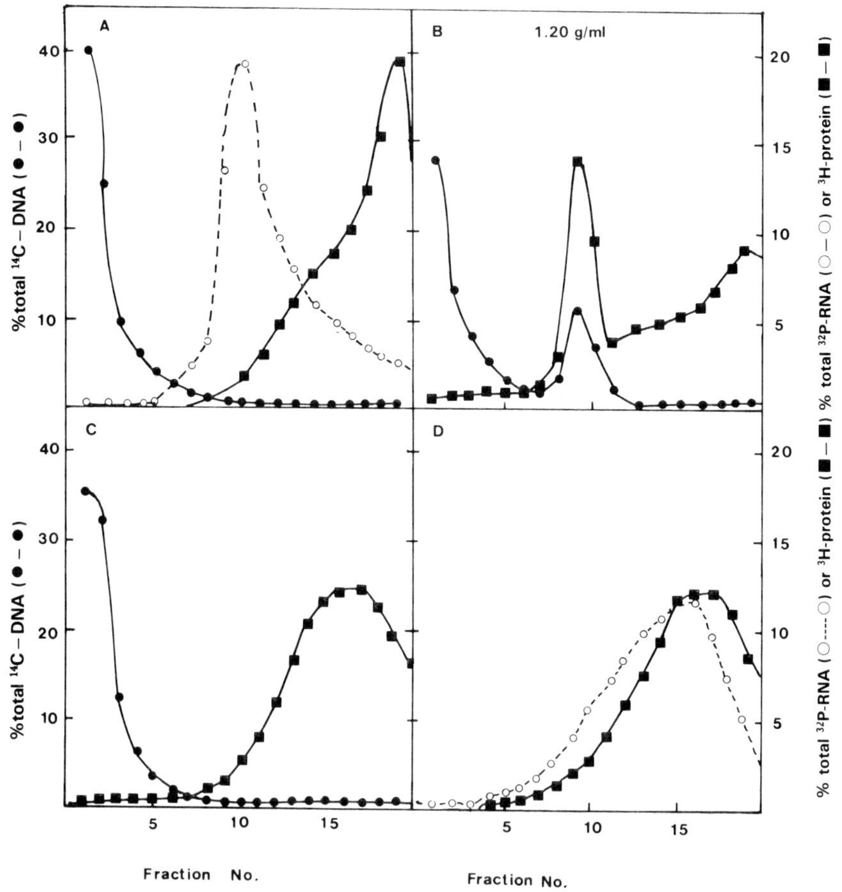

Figure 11. Effect of RNA on the interaction of non-histone proteins with mouse DNA. The proteins, rRNA and DNA were either (A) centrifuged separately or alternatively reconstitution mixtures were centrifuged in 0.14 M NaCl containing (B) DNA (75 μg) and proteins (200 μg); (C) DNA (75 μg) together with 200 μg each of the proteins and unlabelled hnRNA and (D) unlabelled DNA (75 μg) and 200 μg each of labelled proteins and rRNA. The distributions of radioactive DNA (●——●), RNA (○---○) and protein (■——■) were determined in each case. Data from ref. 14.

5.3. Interaction Between Nuclear Non-histone Proteins and DNA

5.3.1. *Introduction*

The nuclear non-histone proteins are distinct from the histones in that they are a very heterogeneous group of proteins, with regard not only to function but also to their physical properties. The proteins have molecular weights ranging from 10 000 to greater than 200 000 and with isoelectric points between pH 3 and pH 10. The basic proteins will, of course, like histones, tend to form complexes with negatively charged nucleic acids. However, of greater interest are those proteins that bind to DNA in a species specific manner since it is generally accepted that such proteins include the proteins that regulate gene expression.

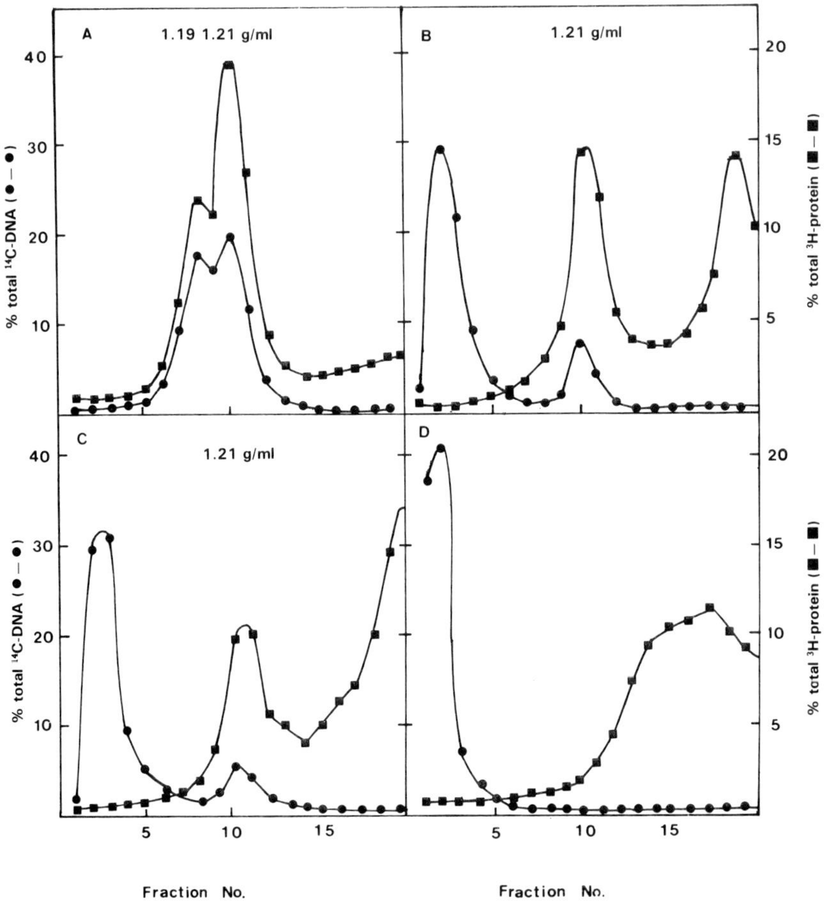

Figure 12. The effect of ionic strength on the binding of non-histone proteins to mouse DNA. The proteins were reconstituted to DNA from 2 M NaCl, 5 M urea into a solution containing either (A) no salt; (B) 0.14 M NaCl; (C) 0.2 M NaCl and (D) 0.3 M NaCl and separated on metrizamide gradients as described in the text. The gradients were fractionated and the density and distribution of DNA (●——●) and protein (■——■) in each fraction were determined. The density of each complex in g/ml is shown in each case. Data from ref. 14.

5.3.2. *Loading of Gradients*

The majority of nuclear non-histone proteins, particularly under conditions of selective binding, do not form stable complexes with DNA. Hence, in these circumstances, it is important to be able to separate the non-bound proteins from the DNA/protein complexes. The problem is that many of the proteins have low molecular weights and hence, take a long while to reach their isopycnic positions if the proteins are originally dispersed throughout the gradient solution (*Figure 10A*). The way to overcome this problem is to load the sample into the bottom of the gradient, then during centrifugation free DNA and DNA/protein complexes float upwards while the proteins remain in the loading zone at the bottom of the gradient (*Figure 10B*). The key to the successful application of this technique is to

Table 8. Experimental Protocol for the Binding of Non-histone Proteins to DNA.

1. *Preparation of the proteins and DNA*
   The proteins should be free of any contamination of RNA or DNA; this can usually be achieved by isopycnic centrifugation in CsCl/urea gradients (19). The DNA should be free of RNA and single-stranded DNA; isolation on hydroxyapatite (23) usually gives DNA of sufficient purity. In some cases it may be appropriate to size the DNA carefully before use or to use specific restriction fragments of defined sequence and size. The DNA and proteins are mixed in a solution containing 2 M NaCl, 5 M urea, 1 mM DTT, 0.1 mM EDTA and 10 mM Tris-HCl (pH 7.5).
2. *Dialysis procedure*
   The exact dialysis procedure has been developed on an empirical basis from studies of chromatin reconstitution. For this procedure first remove the salt to the desired concentration (*Table 7*) and then remove the urea. The exact final concentration of salt depends on the nature of the proteins and their affinity for DNA.
3. *Loading and centrifugation of gradients*
   The sample is mixed with a metrizamide solution of the same ionic composition to give a final concentration of 38% (w/v) metrizamide (Nyegaard & Co., Oslo, Norway). Underlayer 1 ml of the sample under 4 ml of 35% (w/v) metrizamide of the same ionic composition in a 10 ml centrifuge tube and centrifuge the gradients in a MSE 10 x 10 ml or equivalent fixed-angle (21°) rotor at 50 000xg for 40 h at 2°C.
4. *Analysis of gradients*
   Gradients are fractionated by upward displacement. The analysis of the distribution of DNA and proteins is greatly facilitated if both are isotopically labelled (e.g., see *Figures 10–13*).

use DNA large enough to float up to the top of the gradient but small enough such that the density of the DNA is modified significantly when proteins are attached to it.

5.3.3. *Factors that Affect the Interaction of Proteins with DNA*

(i) *Proteins.* Different types of nuclear proteins bind with different affinities to DNA. Hence, for example, proteins easily extracted from nuclei generally, as might be expected, bind more weakly to DNA. In addition, the methods used to prepare the proteins can also affect their DNA-binding activity. For the studies described here the proteins were prepared by the hydroxyapatite procedure (17) which is one of the more gentle isolation procedures; salt extracted proteins also appear to retain a number of their functional activities (18). In contrast, procedures which involve a high degree of denaturation of the proteins, for example, extraction with SDS solutions, are likely to produce much less active protein fractions.

(ii) *Nucleic acid contamination.* Most methods of preparing nuclear proteins yield protein fractions which are contaminated to a greater or lesser extent with nucleic acids, particularly RNA (17). The presence of RNA inhibits the interaction of proteins with the DNA apparently as a result of its preferential binding to the proteins (*Figure 11*). The proteins also seem to bind preferentially to single-stranded DNA in the same way as they do to RNA. Hence, in preparing protein fractions, care should be taken to ensure that the contamination with nucleic acids is minimal; this can be done by purifying the proteins by isopycnic centrifugation on urea-CsCl gradients (19). The DNA used for binding studies should be treated

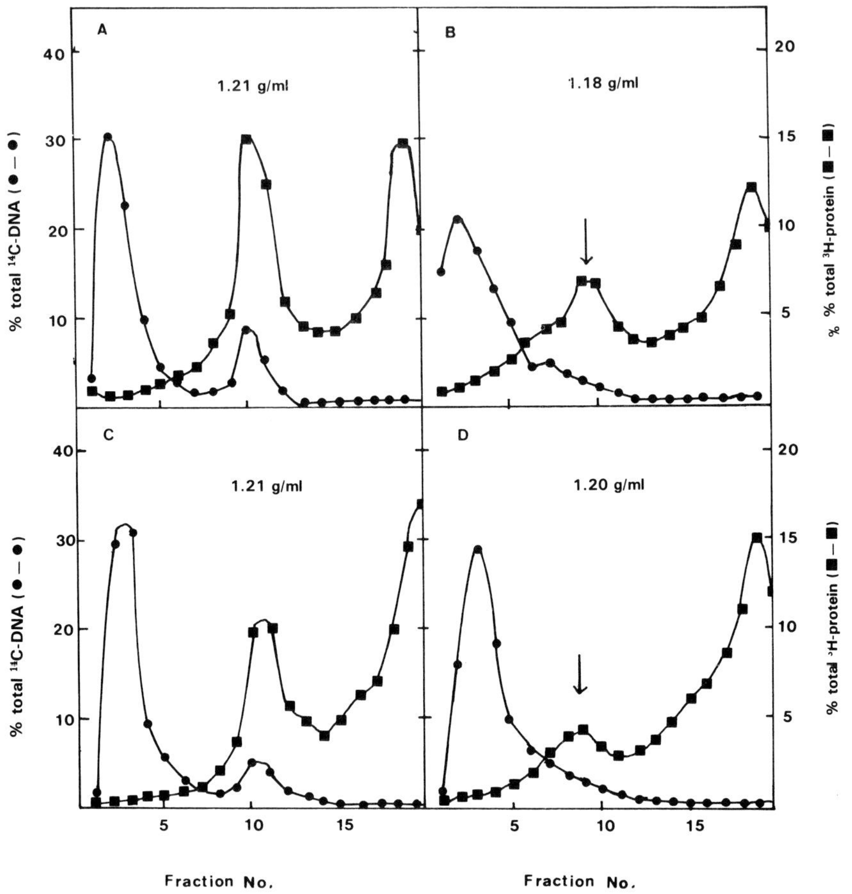

Figure 13. The specificity of the binding of the chromatin non-histone proteins to homologous and bacterial DNA. The proteins were reconstituted to either homologous mouse DNA (A, C) or *E. coli* DNA (B, D) and the complexes were fractionated in gradients containing either 0.14 M NaCl (A, B) or 0.2 M NaCl (C, D). The distribution of DNA (●——●) and protein (■——■) were determined as described in ref. 14. The densities of the complexes in g/ml are shown in each case. Data from ref. 14.

with S1 nuclease to remove any single-stranded fragments of DNA.

(iii) *Ionic environment.* When chromatin non-histone proteins are either mixed with DNA or reconstituted to DNA by dialysis procedures into dilute buffer nearly all of the proteins are bound to all of the DNA (*Figure 12A*). In contrast, when the mixture of DNA and proteins is dialysed into 0.14 M NaCl the proteins bind co-operatively to the DNA; the protein-associated DNA is clearly separated from the non-complexed DNA (*Figure 12B*). A similar pattern of binding is observed when the proteins are dialysed into 0.2 M NaCl (*Figure 12C*), while in 0.3 M NaCl no stable complexes are formed (*Figure 12D*). Depending on the type of proteins under study the stability of the DNA/protein complexes may vary.

5.3.4. *Binding of Nuclear Non-histone Proteins to DNA in a Species Specific Manner*

In order to obtain species-specific binding of proteins to DNA it is necessary to use conditions which are as stringent as possible while still allowing the formation of stable complexes. The protocol that should be used is essentially the same as that given in *Table 8*.

An example of the kind of specificity obtainable is shown in *Figure 13*. The fraction of nuclear non-histone proteins was found to bind to homologous DNA in both 0.14 M NaCl and in 0.2 M NaCl (*Figure 13*). However, when the interaction of these proteins with bacterial DNA is studied, it can be seen that in 0.14 M NaCl less DNA is associated with protein while in 0.2 M NaCl there is no visible peak of DNA complexed with protein. Interestingly, the peak of protein present in 0.2 M NaCl (*Figure 13D*) suggests that the proteins do bind to the *E. coli* DNA but that the complex is not stable and dissociates during centrifugation.

This example is an illustration of the type of study that can be carried out using either metrizamide or Nycodenz gradients to study the interaction of macromolecules. In other situations there may be other, more appropriate, protocols for studying the interaction of proteins and DNA.

6. REFERENCES

1. Birnie,G.D., MacPhail,E. and Rickwood,D. (1974) *Nucleic Acids Res.*, **1**, 919.
2. Birnie,G.D. (1978) in *Centrifugal Separations in Molecular and Cell Biology* (Birnie,G.D. and Rickwood,D. eds.) Butterworths, London and Boston, p. 169.
3. Roussev,G. and Tsanev,R. (1976) *Nucleic Acids Res.*, **3**, 697.
4. Rickwood,D. (1976) in *Biological Separations in Iodinated Density Gradient Media* (Rickwood,D. ed.) IRL Press Ltd., Oxford and Washington, p. 27.
5. Hanansek-Walaszek,M., Walaszek,Z. and Chorazy,M. (1981) *Mol. Biol. Rep.*, **22**, 57.
6. Rickwood,D., Hell,A., Birnie,G.D. and Gilhuus-Moe,C.C. (1974) *Biochim. Biophys. Acta*, **342**, 367.
7. Huttermann,A. and Wendlberger-Schieweg,G. (1976) *Biochim. Biophys. Acta*, **453**, 176.
8. Hartman,G.C., Black,N., Sinclair,R. and Hinton,R.H. (1974) in *Methodological Developments in Biochemistry* (Reid,E. ed.) Longman, London, Vol. **4**, p. 93.
9. Bertoni,J.M. (1981) *J. Neurochem.*, **37**, 1523.
10. Bertoni,J.M. and Steinman,C.G. (1982) *Neurology*, **32**, 323.
11. Hüttermann,A. and Guntermann,U: (1975) *Anal. Biochem.*, **64**, 360.
12. Guenard,D., Morange,M. and Buc,H. (1977) *FEBS Lett.*, **76**, 262.
13. Rickwood,D., Birnie,G.D. and MacGillivray,A.J. (1975) *Nucleic Acids Res.*, **2**, 723.
14. Rickwood,D. and MacGillivray,A.J. (1977) *Expl. Cell Res.*, **104**, 287.
15. Woodcock,C.L.F. (1977) *Science*, **195**, 1350.
16. Birnie,G.D., Rickwood,D. and Hell,A. (1973) *Biochim. Biophys. Acta*, **331**, 283.
17. Rickwood,D. and MacGillivray,A.J. (1975) *Eur. J. Biochem.*, **51**, 593.
18. Wang,T.Y. and Kostraba,N.C. (1977) in *Methods in Cell Biology* (Stein,G., Stein,J. and Kleinsmith,L.J. eds.) Academic Press Inc., New York, San Francisco and London, Vol. **16**, p. 317.
19. Rickwood,D., MacGillivray,A.J. and Whish,W.J.D. (1977) *Eur. J. Biochem.*, **79**, 589.
20. Wolf,A.V. and Brown,M.G. (1971) in *Handbook of Chemistry and Physics* 51st Edn. (Weast, R.C. ed.) Chemical Rubber Co., Cleveland, p. D 176.
21. Cohen,G. and Eisenberg,H. (1968) *Biopolymers*, **6**, 1077.
22. Johns,E.W. (1977) in *Methods in Cell Biology* (Stein,G., Stein,J. and Kleinsmith,L.J. eds.) Academic Press Inc., New York, San Francisco and London, Vol. **16**, p. 183.
23. Hell,A., Birnie,G.D., Slimming,T.K. and Paul,J. (1972) *Anal. Biochem.*, **48**, 369.
24. Ford,T.C., Rickwood,D. and Graham,J. (1983) *Anal. Biochem.*, **128**, 293.

25. Kozmian,L.I., Gauze,G.G., Golkin,V.I. and Dudnik,I.V. (1978) *Antibiotiki,* **23**, 771.
26. Nakamura,H. (1979) *Shika Kiso Igakkai Zasshi*, **21**, 417.
27. Nakamura,H. (1979) *Shika Kiso Igakki Zasshi,* **21**, 428.
28. Ford,T.C. and Rickwood,D. In preparation.
29. Rzeszowska-Walny,J., Gröbner,S., Filipski,J. and Chorazy,M. (1982) *Acta Biochimica Polonica,* **29**, 289.

CHAPTER 3

Fractionation of Ribonucleoproteins from Eukaryotes and Prokaryotes

J.F. HOUSSAIS

1. GENERAL INTRODUCTION

Large macromolecular structures, consisting of high molecular weight RNA and specific proteins, are found in all types of cell. In the cytoplasm of eukaryotic cells, polysomes and their components, the two ribosomal subunits and the messenger ribonucleoproteins (mRNP), are functional parts of the protein synthesising machinery. In the nucleus of these cells are found the heterogeneous ribonucleoproteins (hnRNP) consisting of the primary transcripts of the genome, the nuclear RNA maturation products and the premessenger RNA combined with specific proteins and associated with other nuclear structures.

Structural and functional studies of these macromolecular complexes *in vitro*, require convenient methods of isolation, separation and characterisation. The method used most frequently is fractionation on the basis of size, by rate-zonal centrifugation in sucrose gradients. The sedimentation rate can be measured and the s-values of the individual ribonucleoproteins can be calculated (1). Polysomes sediment as multimeric structures larger than 80S, ribosomes at 30S or 70S, large ribosomal subunits at 60S or 50S and small ribosomal subunits at 40S or 30S, depending upon their eukaryotic or prokaryotic origin *(Table 1)*. The hnRNP particles sediment as heterogeneous structures ranging from 30S to more than 200S. In most cases, rate-zonal sedimentation methods cannot separate the mRNP from the ribosomal subunits since both types of particles cosediment. An alternative method is to separate the RNP particles on the basis of density in appropriate media. Isopycnic centrifugation in density gradients of salts of alkali metals (e.g. caesium chloride) has been shown to be useful, but, to prevent dissociation of the RNA and protein components in solutions of such high ionic strength, it is necessary to stabilise the RNP structure by prior fixation with formaldehyde or glutaraldehyde. This fixation hinders any further study of the separated particles. In contrast, metrizamide or Nycodenz, which are both completely non-ionic compounds of low viscosity over the usual density ranges, appear to be useful alternatives for separating cellular ribonucleoproteins.

This chapter presents a detailed analysis of the usefulness of these two non-ionic iodinated media, metrizamide and Nycodenz, for separating polysomes, their components, and the hnRNP particles. Original observations, obtained with mammalian cells in culture, will be presented as examples. To comple-

Table 1. The Sedimentation Constants of the Ribonucleoprotein Particles of Cells.

Type of particles	*Prokaryotic*	*Eukaryotic*	*Sedimentation pattern*	*Ref.*
Polysomes	100-300S	150-400S	Polydispersed with discrete peaks ahead of the ribosome peak	33-36
Ribosomes	70S	80S	Discrete peak	37
Large ribosomal subunits	50S	60S	Discrete peak	37
Small ribosomal subunits	30S	40S	Discrete peak	37
Messenger ribonucleoprotein particles (polysomal mRNP)		10 to 70S	Polydispersed	11,12,38
Free cytoplasmic messenger nucleoprotein particles (mRNP)	None	10 to 80S	Polydispersed	4,5,38
Heterogeneous nuclear ribonucleoprotein particles (hnRNP)	None	30 to >200S	Polydispersed	25

The relationship between size and sedimentation constants of the ribonucleoproteins as analysed by zone velocity sedimentation in sucrose gradients. For the methodology used, see refs. 1 and 32. In prokaryotic cells, synthesis of mRNA (transcription) and synthesis of proteins on polysomes (translation) occur almost concomitantly; there is no hnRNP particle, and no mRNP particles have been individualised as such. In contrast, in eukaryotic cells, transcription and translation occur in two distinct cell compartments, the nucleus and the cytoplasm. The synthesis and maturation of mRNA occur in the nucleus and the synthesis of ribosomal subunits in the nucleolus, whereas the synthesis of proteins occurs on the polysomes in the cytoplasm. For clarity the sedimentation constants of the ribosomal particles of subcellular organelles (mitochondria, chloroplasts) have not been included in this table.

ment them, data obtained from yeast and *E. coli* ribosomes, will also be quoted. In this chapter the methodologies and example results will be given in detail, and practical points will be discussed, so as to enable the reader, either to reproduce the technique described or, starting from the data presented, to modify the experimental conditions to suit similar types of fractionation.

2. SEPARATION OF POLYSOMES AND THEIR COMPONENTS

2.1. Introduction

To facilitate the evaluation of iodinated density gradients for separating polysomes, the two media, metrizamide and Nycodenz, were studied in parallel in the same experiments. In all of these experiments the following experimental procedures were used. Ethylenediaminetetraacetate (EDTA) was used to dissociate polysomes into ribosomal subunits and mRNP. Ribosomal RNA and proteins were both labelled in order to determine if there is any apparent dissociation of the nucleoprotein complexes. Polysomal RNA was labelled in the presence or absence of low doses of actinomycin D in the culture medium; actinomycin D selectively blocks the synthesis of the nucleolar ribosomal RNA without affecting the synthesis of messenger RNA (2,3), and this differential effect provides a convenient method of following the sedimentation of the mRNP into the gradient. In addition, ribonucleoproteins have been separated on Nycodenz gradients with different density profiles.

2.2. Methods and Experimental Procedures

2.2.1. *Cell Culture*

The mouse cell line Cl1D LM (TK^-) was used as cell material; other types of cells can also be used. The cells were grown in suspension culture at 37°C in Eagle-Dulbecco's medium (GIBCO), supplemented with 10% calf serum (3).

2.2.2. *Method for the Labelling and Harvesting of Cells*

Label the RNA and proteins for 4 h by adding to the culture medium, 0.07 μCi/ml (2.6 kBq/ml) of ^{14}C-uridine (Amersham International plc, 488 mCi/mmol, 18 GBq/mmol) and 0.5 μCi/ml (18.5 kBq/ml) of [4,5-^{3}H]leucine (Amersham International plc, 130 Ci/mmol, 5 TBq/mmol).

If required, add actinomycin D (Calbiochem) to the culture medium half an hour before labelling with radioisotopes, at a concentration of 0.05 μg/ml. Twenty minutes before harvesting the cells, add cycloheximide (Calbiochem) at 0.5 μg/ml (3), to stabilise the polysomal structure during isolation. Centrifuge the cells at 3000xg for 5 min, and wash the cells with isotonic phosphate buffered saline. Pellets of 1 – 2 ml of packed cells are collected by centrifugation and stored at – 70°C until required.

2.2.3. *Method for the Isolation and Purification of Polysomes*

Resuspend the packed cells in 20 volumes of the homogenisation solution (50 mM sucrose, 200 mM ammonium chloride, 7 mM magnesium acetate, 1 mM dithiothreitol, and 20 mM Tris-HCl, pH 7.6). Lyse the cells by adding the nonionic detergent Nonidet P40 to a final concentration of 0.5% followed by a few strokes by hand in a glass-Teflon homogeniser. Pellet the nuclei and cellular fragments by centrifugation at 10 000xg for 10 min at 4°C. Load the supernatant onto a two-step discontinuous gradient of 1.8 M and 1.0 M sucrose both in 100 mM ammonium chloride, 5 mM magnesium acetate, 1 mM dithiothreitol (DTT), 20 mM Tris-HCl (pH 7.6), in Beckman SW27 or equivalent centrifuge tubes. Centrifuge the gradients in a SW27 rotor or equivalent for 18 h at 26 000 rev/min (98 000xg) at 5°C (3). At the end of the run, discard the supernatants and layers of sucrose and carefully wipe out the inside of each tube being careful not to disturb the pellet of polysomes. Gently resuspend the pelleted polysomes in 100 mM ammonium chloride, 5 mM magnesium acetate, 1 mM DTT and 20 mM Tris-HCl (pH 7.6). To dissociate the polysomes into ribosomal subunits and mRNP, resuspend the packed polysomes in the same solution but with the magnesium ions replaced by 5 mM EDTA. Determine the amount of polysomal material by measurement of the optical density at 260 nm; assume that one optical density unit corresponds to 60 μg/ml. If polysomes are not to be used immediately, store them at – 70°C as packed polysomes.

2.2.4. *Preparation, Centrifugation and Analysis of Gradients of Iodinated Density-Gradient Media*

Density gradients may be preformed or generated by the centrifugal field during centrifugation. The use of self-forming gradients requires careful choice of the initial density and generally requires longer centrifugation times (see Section 4.2 of chapter 1) but it is easier to prepare the starting solutions of the iodinated compounds. It is this technique which has been used both with metrizamide and Nycodenz, and which is now described in detail.

First decide on the buffered solution into which the iodinated compound will be dissolved. For this study, the following solution has been used throughout: 20 mM triethanolamine-HCl (pH 7.6), 25 mM KCl, 2.5 mM magnesium acetate. The magnesium ion concentration appears to be critical for a good separation of the cytoplasmic RNP particles (4-6). The best results are obtained using 2 – 4 mM magnesium ions. Since, in the absence of magnesium ions banding artifacts may occur (4,7), all gradients contained 2 mM magnesium ions; it is known that no reassociation of ribosomal subunits occurs after mammalian polysomes have been dissociated with EDTA.

Prepare the stock solutions of the iodinated media before centrifugation (this usually takes about 1 – 2 h); the volume needed is dependent upon the type of rotor used. Choose the initial concentration of gradient solute so that the centrifugation conditions used generate the correct density range and gradient profile needed for the separation. For example, gradients of 36% (w/v) metrizamide allow the separation of ribosomal subunits and mRNP, but not of polysomes which migrate to the bottom of the tube (see Section 2.3.1.). To band polysomes in metrizamide gradients the metrizamide concentration has to be increased to at least 45% (w/v), and in the case of Nycodenz, to 49 – 50% (w/v) using the centrifugation conditions described here (see Section 2.3.2.). By using either a solution of uniform concentration, or several layers of decreasing concentrations, it is possible to modify the shape of the final density profile; a solution of 50% (w/v) Nycodenz gives a sigmoid shaped density profile, with a significant plateau in the middle of the gradient (see *Figures 2A* and *3A*) while two layers of 35% and 50% (w/v) of equal volume give, at the end of a 65 h run, a nearly linear density profile (see *Figures 2B* and *3B*).

Using this very simple method, it is possible to fractionate ribosomes and polysomes as follows. Prepare three solutions:

Solution	*Weight of gradient solute*	*Add buffered solution up to a final volume of*	*Final concentration (w/v)*
Solution 1	7.50g Nycodenz	15.0 ml	50%
Solution 2	1.75g Nycodenz	5.0 ml	35%
Solution 3	4.50g metrizamide	10.0 ml	45%

Dissolve the iodinated compound and adjust to volume. Metrizamide dissolves easily while Nycodenz takes a little longer time, and it may be useful to warm the solution in a water-bath up to 50°C.

A typical separation of polysomes can be carried out as follows. Into two Beckman centrifuge tubes for the SW50.1 rotor or equivalent, add 5.0 ml of solution 1 ("50% Nycodenz"), add 5.0 ml of solution 3 into two other tubes ("45% metrizamide"). Then into the last two tubes, add first 2.5 ml of solution 1 per tube, and carefully layer 2.5 ml of solution 2 ("Nycodenz 35 – 50%") (*Figures 2* and *3*).

Layer onto each gradient separate samples of polysomes or ribosomal subunits at about 7 $O.D._{260nm}$ units per tube, in 0.5 ml of the appropriate buffer (plus or minus EDTA depending on the sample). Centrifuge the gradients in a SW50.1 rotor or equivalent for 65 h at 4°C and 30 000 rev/min (95 000xg). It is possible to use other centrifugation conditions, different length tubes, other swing-out or fixed-angle rotors, etc. However, it is then necessary to adjust the concentration of the gradient solute together with the centrifugation time (8). At the end of the centrifugation run, fractionate the gradients by any of usual methods. For simplicity, the author has found it convenient to use an Isco fractionator slightly modified so that gentle air-pressure at the top of the gradient pushes the gradient through a fine bore hypodermic needle pushed into the bottom of the tube. Routinely the gradients are fractionated into 5 drop fractions.

The collected fractions can be stored at – 25°C, or processed immediately to obtain information on the density profile of the gradient and the banding pattern of the RNP particles. Although the density of each fraction could be obtained by direct weighing of a known volume, it is more convenient to measure the refractive index (5 – 20 μl/fraction) using an Abbé refractometer, and to determine the density either by using the following equations:

Metrizamide:

$$\text{density (g/ml)} = 3.453\eta_{20°C} - 3.601 \qquad (9)$$

Nycodenz:

$$\text{density (g/ml)} = 3.242\eta_{20°C} - 3.322 \qquad (10)$$

alternatively use a graph of a linear plot of the relationship between the refractive index and density.

To assess the banding of the RNP particles, two methods can be used, either assay of the RNA by the orcinol reaction (see Appendix 1) after acid precipitation of each fraction with 10% trichloroacetic acid (TCA) or by measuring the radioactivity of each fraction, after TCA precipitation in presence of 100 μg of serum albumin as coprecipitant, and filtration on glass-fibre filters (GF/C). This second method was used in these experiments.

2.3. Isopycnic Banding of Mammalian Polysomes, Ribosomal Subunits and mRNP

2.3.1. *Differential Banding in Gradients of 36% (w/v) Metrizamide*

Figure 1 demonstrates that polysomes, ribosomal subunits and mRNP particles can be clearly identified in iodinated density-gradient media. Polysomal RNA can be labelled in the presence or absence of actinomycin D. In both cases, the undissociated polysomes and ribosomes have the same radioactivity profile as the control *(Figure 1A)*. After EDTA dissociation, the radioactivity profile of the control *(Figure 1B)* shifts to a major peak corresponding to the ribosomal subunits, and a minor peak with a density of 1.18 g/ml is present.

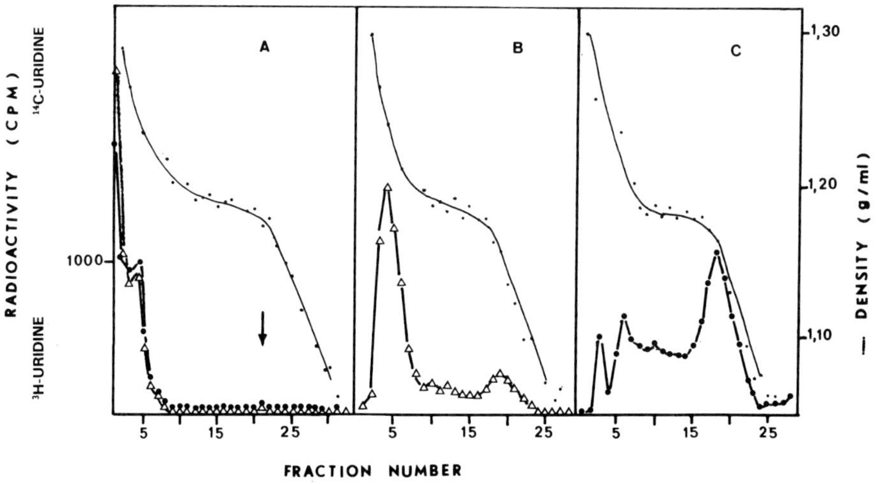

Figure 1. The isopycnic banding of polysomes, ribosomal subunits and polysomal messenger ribonucleoproteins, in 36% (w/v) metrizamide self-forming gradients. The polysomal RNA was labelled with ^{14}C-uridine in presence of 0.05 μg/ml of actinomycin D (●——●), or with ^{3}H-uridine in absence of actinomycin D (△——△). Panel A: radioactivity profile of undissociated polysomes, labelled in presence or absence of actinomycin D. Panel B: radioactivity profile of the EDTA-dissociated polysomes of the control, labelled without actinomycin D in the medium. Panel C: radioactivity profile of the EDTA-dissociated polysomes labelled in presence of actinomycin D. The labelling time was 4 h. The self-forming gradients were obtained by centrifugation at 95 000xg at 4°C for 65 h in the SW 50.1 rotor.

When the polysomal RNA is labelled in presence of actinomycin D, which selectively suppresses the synthesis of rRNA, there is very little radioactivity in the ribosomal region of the gradient *(Figure 1C)*. The residual radioactivity which is known to be due to the messenger RNA in the polysomal mRNP (11,12), gives a major peak at density 1.18 g/ml, which extends heterogeneously between 1.18 and 1.20 g/ml *(Figure 1C)*.

2.3.2. *Separation of Polysomes and EDTA Dissociated Polysomes in 50% (w/v) Nycodenz, 35 – 50% (w/v) Nycodenz, and in 45% (w/v) Metrizamide*

Figure 2 shows the banding pattern of undissociated control polysomes, that is, labelled in absence of actinomycin D *(Figures 2, A_1,B_1,C_1)*, or control EDTA-dissociated polysomes *(Figures 2, A_2,B_2,C_2)*, in 50% (w/v) Nycodenz *(Figures 2 A_1,A_2)*, 35 – 50% (w/v) Nycodenz *(Figures 2B_1,B_2)* and 45% (w/v) metrizamide *(Figures 2 C_1,C_2)*. The general banding pattern appears similar in both media. As in *Figure 1*, a density shift occurs when the polysomes are dissociated by EDTA into their subunits. The radioactivity in the mRNP region of the gradients is low as expected, increasing slightly after dissociation of the polysomes. These data parallel the results shown in *Figures 1A* and *1B*. However, in the case of *Figure 2*, three observations, which will be discussed later, should be noted:

(i) more radioactive material appears in the 1.18 – 1.20 g/ml density range,

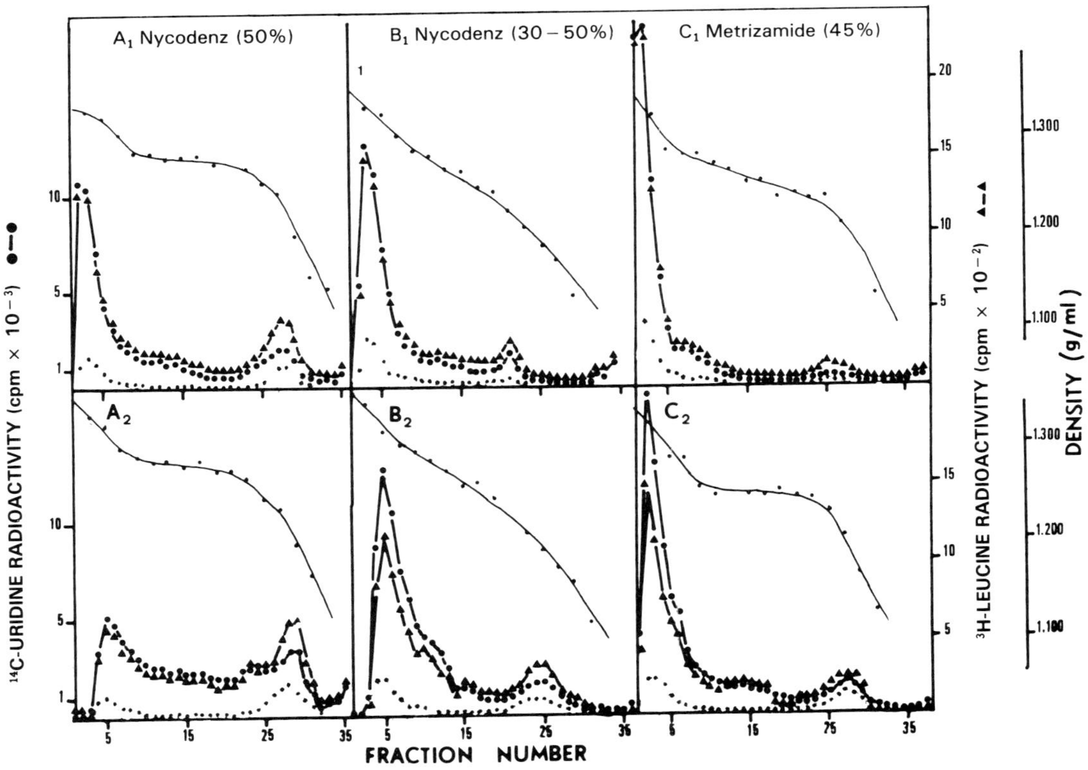

Figure 2. Comparative density profiles of polysomes and EDTA-dissociated polysomes in metrizamide and Nycodenz self-forming gradients. Panels A_1,A_2: 50% (w/v) Nycodenz. Panels B_1,B_2: half 50%, half 35% (w/v) Nycodenz, in two layers. Panels C_1,C_2: 45% (w/v) metrizamide. The buffer and the conditions of centrifugation were as in *Figure 1* and as described in Methods and Experimental Procedures: Polysomal RNA were labelled with ^{14}C-uridine (●——●). Polysomal proteins were co-labelled with ^{3}H-leucine (▲——▲). The labelling time was 4 h. A_1,B_1,C_1: density profile of undissociated polysomes. A_2,B_2,C_2: density profile of EDTA-dissociated polysomes. The lower curve of each graph (-----) represents the RNA radioactivity profile of a same amount of polysomes and EDTA-dissociated polysomes, which have been labelled for 4 h, in presence of actinomycin D. These curves are enlarged in *Figure 3*.

when Nycodenz is used (compare *Figures 2* A_1,B_1 with C_1);
(ii) the density profile obtained when Nycodenz is 35 – 50% (w/v) seems better for separating the ribosomal subunits than 50% (w/v) Nycodenz (compare with *Figures* 2A_2 to B_2);
(iii) the ribosomal subunits are better separated in 35 – 50% (w/v) Nycodenz than 45% (w/v) metrizamide.

The lower curve on each panel of *Figure 2* shows, using the same scale, and for the same amount of material loaded onto each gradient, the radioactivity profiles of polysomal RNA labelled in presence of actinomycin D. After EDTA dissociation, one finds very little residual incorporation into the ribosomal peaks, compared with the control (*Figures 2* B_2,C_2). Figure 3 shows the lower curves of in *Figure 2* drawn on a larger scale. When the polysomes are undissociated, most of the radioactivity is found in the polysome peak (*Figures 3* B_1,C_1). After EDTA dissociation, a density shift to the region of ribosomal subunits is observed (indicating that in these experiments the actinomycin inhibition of rRNA synthesis was not completely effective), and also a relatively high radioactivity in the mRNP region of the gradients at density 1.18 – 1.22 g/ml (*Figures* 3B_2,C_2). This technique of labelling in presence of actinomycin D, provides a sensitive method of following what happens to the mRNP part of polysomes. If we assume that any accumulation of radioactivity in the mRNP region of the gradients indicates that some degree of polysomes dissociation has occurred, then it should be noted that centrifugation in 50% (w/v) Nycodenz appears to produce a partial dissociation of the polysomes (compare *Figure* 3A_1 with *Figures* 3B_1,C_1, and *Figure* 3A_1 with *Figure* 3B_1). In contrast, the integrity of polysomes seems better preserved in 35 – 50% (w/v) Nycodenz and 45% (w/v) metrizamide gradients.

2.3.3. *Migration of Purified Large and Small Ribosomal Subunits in Nycodenz and Metrizamide Gradients*

The ribosomal subunits were purified, in order to compare more precisely their banding in metrizamide and Nycodenz gradients. Labelled polysomes were dissociated by EDTA, and the subunits sedimented through a 15 – 35% sucrose gradient containing 1 mM EDTA (3). Large and small ribosomal subunit fractions were pooled as indicated on *Figure 4*, and layered directly onto the 50%, 35 – 50% (w/v) Nycodenz and 45% (w/v) metrizamide. It should be remembered that sucrose gradients only separate particles on the basis of size and while the two ribosomal subunits are well resolved (*Figure 4*) some mRNP comigrates with the ribosomal subunits (11,12). So ribosomal subunits must be expected to be contaminated to some extent by mRNP.

Figure 5 shows the banding of the large ribosomal subunits. They band as a major peak in 35 – 50% (w/v) Nycodenz and 45% (w/v) metrizamide, with a little material in the mRNP region of the gradient (*Figures 5B* and *C*). In 50% (w/v) Nycodenz, the banding pattern of the large subunits appears more heterogeneous, this is because the density of the plateau is too close to that of the particles.

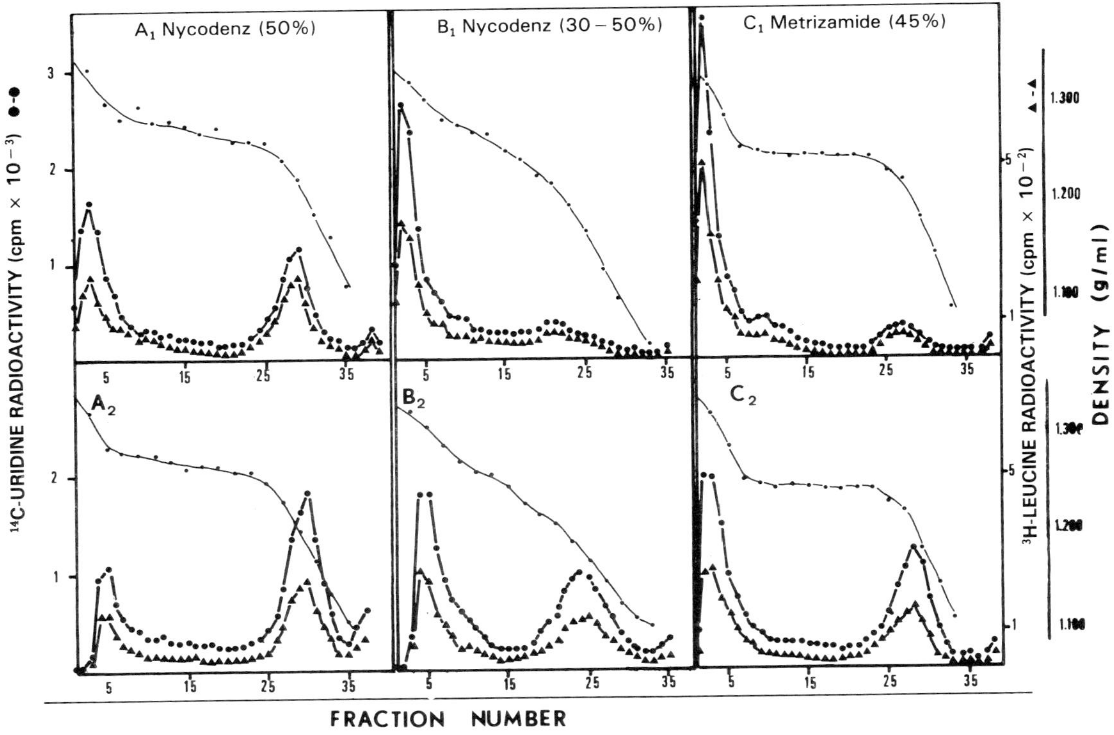

Figure 3. Comparative density profiles of polysomes and EDTA-dissociated polysomes, in metrizamide and Nycodenz self-forming gradients, after 4 hours labelling of RNA, in presence of actinomycin D. Low doses of actinomycin D (0.05 μg/ml of medium) were used. The concentrations of the starting solutions of metrizamide and Nycodenz, and the meaning of symbols are as in *Figure 2*. (●——●) Radioactivity of RNA. (▲——▲) Radioactivity of proteins.

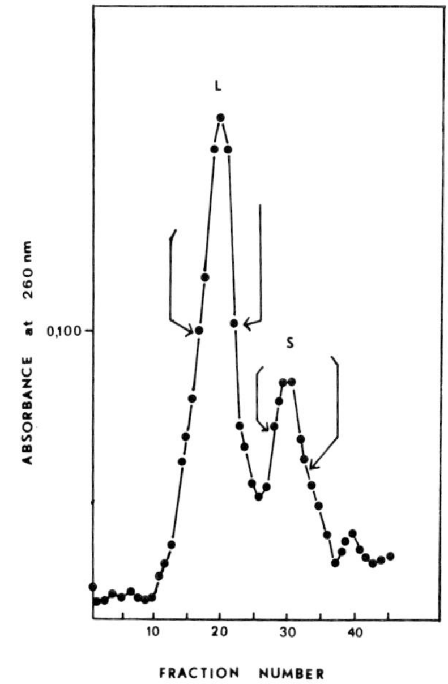

Figure 4. Fractionation and purification of the large and small ribosomal subunit fractions in 15–30% (w/v) sucrose gradients, after EDTA-dissociation of polysomes. Gradients of 15-30% (w/v) sucrose were made up in 1 mM EDTA, 0.025 M KCl and 0.05 M Tris-HCl (pH 7.6) and 23 $O.D._{260nm}$ units of polysomes dissolved in 1.5 ml of the same solution, were layered onto the sucrose gradients in Beckman tubes for the SW 27 rotor, and centrifuged at 23 000 rev/min (75 000xg) for 16 h (3). Fractions were collected and the absorbance read at 260 nm. The fractions corresponding to the large subunit (L) and to the small subunits (S) were pooled as indicated in the Figure.

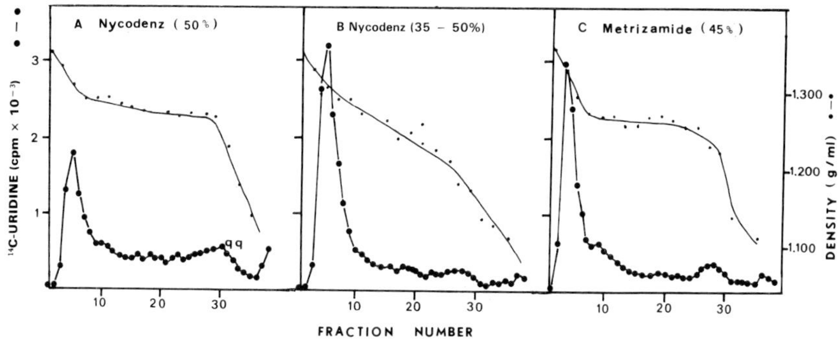

Figure 5. The density profiles of the purified large ribosomal sybunits layered onto (A) 50% (w/v) Nycodenz, (B) 35–50% (w/v) Nycodenz, and (C) 45% (w/v) metrizamide, and centrifuged under conditions similar to those summarised in *Figure 2.*

Figure 6 shows the banding pattern of the material isolated with the small ribosomal subunits from the sucrose gradient (see *Figure 4*). The results show the following:

(i) the small subunits are apparently more contaminated by mRNP than the

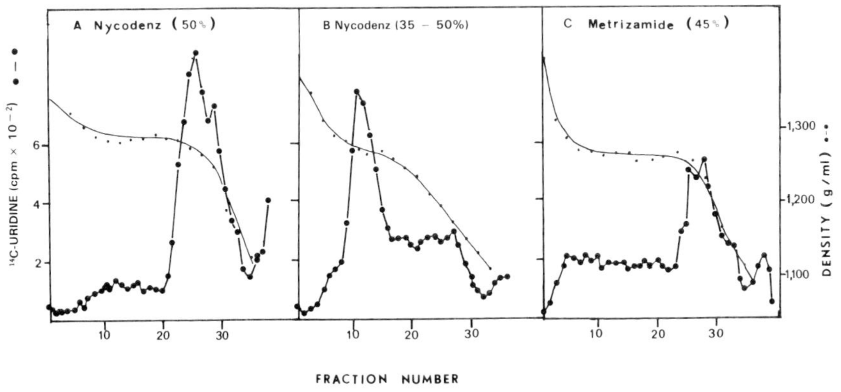

Figure 6. The density profiles of the small ribosomal subunits and the associated mRNP, in (A) 50% (w/v) Nycodenz, (B) 35 – 50% (w/v) Nycodenz, and (C) 45% (w/v) metrizamide, under conditions similar to those of *Figures 2* and *5*.

large subunits, with the mRNP banding as an heterogeneous peak in the range 1.18 – 1.25 g/ml region of the gradients *(Figure 6B)*;
(ii) the small shoulder in the high density region, suggests that this fraction is contaminated by a small amount of large ribosomal subunit contaminant;
(iii) the small subunits migrate as a major peak in the Nycodenz gradients *(Figures 6A,B)* however they are not well resolved in the 45% (w/v) metrizamide gradients; possibly because the density of the plateau is too high.

2.4. Advantages and Disadvantages of Using Iodinated Gradient Media for the Separation of Mammalian Polysomes and Their Components

2.4.1. *Determination of the Best Gradient Profile*

When using self-forming density gradients and constant centrifugation conditions (rotor, time and speed), two parameters are important for separating the RNP, as illustrated by the previous results, the shape of the density profile and the density of the plateau. Both parameters are dependent upon the concentration of the iodinated compound in the starting solution, which has to be adjusted as we have seen to suit the apparent density of the RNP to be separated. With a 50% (w/v) Nycodenz and 45% (w/v) metrizamide, we get an accentuated sigmoidal density curve, with a large plateau *(Figures 2, 3, 5,* and *6)*. While this unusual shape may give a better resolution if properly adapted, for instance, to separate the ribosomal subunits and the mRNP, it may be of no use if the density plateau is too close to that of the density of the RNP particles. We have seen examples for the large subunits in 50% (w/v) Nycodenz *(Figure 5A)* and for the small subunits in 50% (w/v) Nycodenz and 45% (w/v) metrizamide *(Figures 6A,C)*. One can smooth the density curve to an almost linear shape, simply by preparing a two-step gradient from two solutions of 35% and 50% (w/v) Nycodenz. From the results described in the preceding section, it appears that a gradient with a more linear slope is more suitable for separating a mixture of several ribonucleoproteins of similar densities such as polysomes, large and small ribosomal subunits, and mRNP.

2.4.2. *Apparent Buoyant Densities of Mammalian Polysomes, Ribosomal Subunits and mRNP in the Iodinated Media, Metrizamide and Nycodenz*

Table 2 summarises the estimated values of the apparent densities of polysomes, ribosomal subunits and mRNP, as observed in this work, and as reported for metrizamide in other publications. As shown in *Figures 2* and *3*, polysomes have almost the same density in Nycodenz (1.334 g/ml) and in metrizamide (1.34 g/ml). However, differences occur for the ribosomal subunits. In Nycodenz, the large subunit is well separated from polysomes, and bands at a density of 1.304 g/ml, while in metrizamide, the apparent density of the large subunit remains near that of polysomes at 1.33 g/ml. The small ribosomal subunit bands as a well separated peak at a density of 1.27 g/ml in Nycodenz, while, as we have seen, its density in 45% (w/v) metrizamide can only be roughly estimated in these experiments at 1.26 – 1.27 g/ml. The mRNP particles band over the same density range in both metrizamide and Nycodenz at 1.18 – 1.22 g/ml. The density values obtained in metrizamide, are similar to those found in other studies, where self-forming gradients have been used (4,5); they appear different from when metrizamide gradients have been preformed and shorter centrifugation times used (13).

2.4.3. *Effects of Iodinated Media and of the Centrifugal Field on the Structural Integrity of the Mammalian Polysomes*

It is an unexpected observation that, as reported in this study, polysomes exhibit a tendency to dissociate spontaneously into subunits and mRNP. This is shown, as we have seen, in *Figures 2* and *3*. This effect is minor in 45% (w/v) metrizamide (*Figures* $2C_1$, $3C_1$) and in 35 – 50% (w/v) Nycodenz where the density profile is nearly linear (*Figures* $2B_1$, $3B_1$). However, if the starting concentration of Nycodenz is high (50%), together with the high level of the plateau, polysomes dissociate into subunits and mRNP (*Figures* $2A_1$, $3A_1$). At least three factors might play a role in this phenomenon:

(i) the relatively low magnesium ion concentration used (2 – 2.5 mM);
(ii) an effect of this particular medium at high concentration;
(iii) the forces generated by the centrifugal field.

In fact, in sucrose gradients, pressure induced dissociation of polysomes during ultracentrifugation has been described (14,15). It is possible that these effects might be minimised by using a near linear density gradient containing a slightly increased magnesium ion concentration to stabilise the polysomal structure.

2.4.4. *The Iodinated Media: a Potential Means for Isolating Poly(A)$^+$ and Poly(A)$^-$ Polysomal Messenger Ribonucleoproteins*

Metrizamide gradients have proved to be an efficient method for isolating the cytoplasmic free-mRNP which were found to band at a density of 1.20 – 1.22 g/ml (4,5). From a study involving rat-liver polysomes (13), it was concluded

Table 2. The apparent buoyant densities[a] of mammalian cytoplasmic ribonucleoproteins in Nycodenz and Metrizamide gradients.

Cell material (mammalian cells)	*Polysomes*		*Large subunits*		*Small subunits*		*Polysomal mRNP*		*Cytoplasmic free mRNP*		*Experimental conditions*	*Refs.*
	Nycodenz	*Metrizamide*	*Nycodenz*	*Metrizamide*	*Nycodenz*	*Metrizamide*	*Nycodenz*	*Metrizamide*	*Nycodenz*	*Metrizamide*		
Mouse L cells, Cl1D LM (TK^-)	1.334 ± 0.002	1.34	1.304 ±0.005	1.33	1.270	1.25-1.27	1.15-1.23	1.15-1.23	n.d.[b]	n.d.	Self-forming gradients 2-2.5 mM Mg^{2+} Nycodenz 50% Nycodenz 35-50% metrizamide 45% (w/v)	
Primary cultures of muscle cells of foetal calves		1.35		1.32		1.26		1.23-1.24		1.205	Self-forming gradients 3 mM Mg^{2+} metrizamide 40% (w/v)	4,5
Rat liver		1.295-1.300		1.23		1.20-1.21		1.12-1.23			20-60% (w/v) preformed metrizamide gradients 5 mM Mg^{2+}	13

[a]All values g/ml. [b]Not done. By comparison with the above values, the apparent buoyant density of other eukaryotes and of prokaryotes has been determined:
- yeast 80S ribosomes: 1.32 g/ml, in self-forming 48% (w/v) metrizamide at 3 mM Mg^{2+} (ref. 6).
- *E. coli* 70S ribosomes: 1.33 g/ml, in self-forming 47% (w/v) Nycodenz (ref. 19).

that polysomal mRNP could not be completely separated from the ribosomal subunits in preformed metrizamide gradients, since the mRNP overlapped the density region of the subunits. In contrast, this study demonstrates that individual peaks of mRNP in the 1.18 – 1.22 g/ml density range in both metrizamide and Nycodenz, are well separated by density *(Table 2)* from the ribosomal subunits (*Figure $2B_2$* and *Figures $3B_2,C_2$*). Furthermore, as shown in *Figure 6*, it seems clear that 35 – 50% (w/v) Nycodenz gradients give better resolution than 50% (w/v) Nycodenz and 45% (w/v) metrizamide gradients. Although this point has not been studied further, the results presented in *Figure 3* and *Figure 6B* strongly suggest that linear Nycodenz density gradients might prove a useful method for isolating total poly(A)$^+$ and poly(A)$^-$ polysomal mRNP.

2.5. **Isopycnic Banding of Non-mammalian Ribosomes**

2.5.1. *Banding Properties of Yeast 80S Ribosomes (6)*

Purified, fixed or unfixed 80S ribosomes of yeast, have a buoyant density of 1.32 g/ml in self-forming 48% (w/v) metrizamide gradients containing 3 mM magnesium ions. The ribosomes can be either top or bottom loaded, or mixed throughout the gradient solution. It has also been shown (6) that 40% (w/v) metrizamide slowly interacts with the loosely associated proteins of the 80S ribosomes. From the shift in density of the unfixed particles in CsCl gradients, from 1.52 g/ml to 1.55 g/ml, it can be calculated that about 10% of the ribosomal proteins are in equilibrium with the ribosome *(Figure 7)*. However, this effect does not appear to affect the structural integrity of polysomes and ribosomal subunits in the experiments described *(Figures 2* and *3)* using mammalian polysomes where RNA and proteins have both been labelled. It is known that metrizamide only interacts with proteins not complexed with nucleic acids (14-16). The observations reported for yeast 80S ribosomes, suggest that each experimenter should check the extent of this dissociation effect in each case. The data shown in *Figures 2* and *3* on the polysomal dissociation induced in iodinated media and in the centrifugation field, could be related to the interaction of the media with some of the proteins, making the polysomal structure less stable.

Following the studies on mammalian ribosomes (4,5), the effect of the magnesium concentration on the banding and apparent density of yeast 80S ribosomes has been investigated (6). It appears that changes in conformation and hydration are caused by increasing magnesium concentrations and these are reflected by the dramatic changes of the buoyant density of the 80S ribosomes. The density shifts from 1.22 g/ml in the absence of magnesium ions to 1.32 g/ml at 3 mM magnesium ions. Higher concentrations, 10 mM and 40 mM magnesium ions, lead to a number of apparently discrete denser complexes at 1.33 g/ml, 1.35 g/ml and 1.41 g/ml *(Figure 8)*.

2.5.2. *Banding Properties of E. coli 70S Ribosomes (19)*

E. coli 70S ribosomes, purified by rate-zonal centrifugation on a 10 – 40%

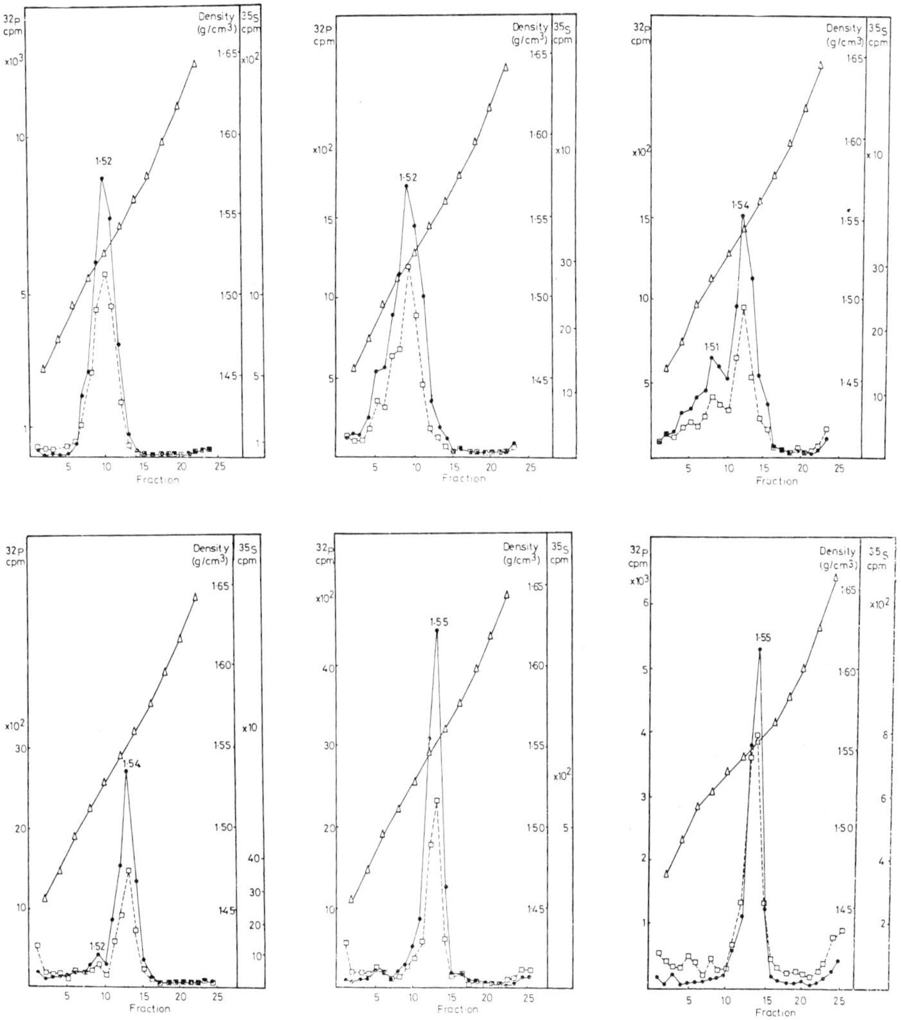

Figure 7. The density shift of yeast ribosomes, in metrizamide. Ribosomes were left in the presence of 40% (w/v) metrizamide for various times up to 24 h, then fixed in 7% (w/v) HCHO for 2 h and centrifuged in CsCl gradients containing 3 mM Mg^{2+}, 1% (w/v) HCHO, 10 mM Hepes-NaOH (pH 7.0 at 100 000xg for 18 h at 4°C). (A) control; (B) ribosomes left in 40% (w/v) metrizamide for 15 min. (C) Ribosomes left in 40% (w/v) metrizamide for 30 min. (D) Ribosomes left in 40% (w/v) metrizamide for 60 min. (E) Ribosomes left in 40% (w/v) metrizamide for 90 min. (F) Ribosomes left in 40% (w/v) metrizamide for 24 h. ●——● ^{32}P; □– – –□ ^{35}S; △——△ density.

(w/v) sucrose gradient, were centrifuged in self-forming Nycodenz density gradients (4 ml of 47% (w/v) Nycodenz, with a 1.0 ml 60% (w/v) Nycodenz cushion) at 75 000xg for 44 h at 5°C in a MSE 10x10 ml aluminium rotor. The gradients contained 10 mM $MgCl_2$, 70 mM KCl, 10 mM Hepes-NaOH (pH 7.4). The 70S ribosomes were centrifuged either unfixed or after fixation with 7% formaldehyde. The banding density of both fixed and unfixed *E. coli* 70S

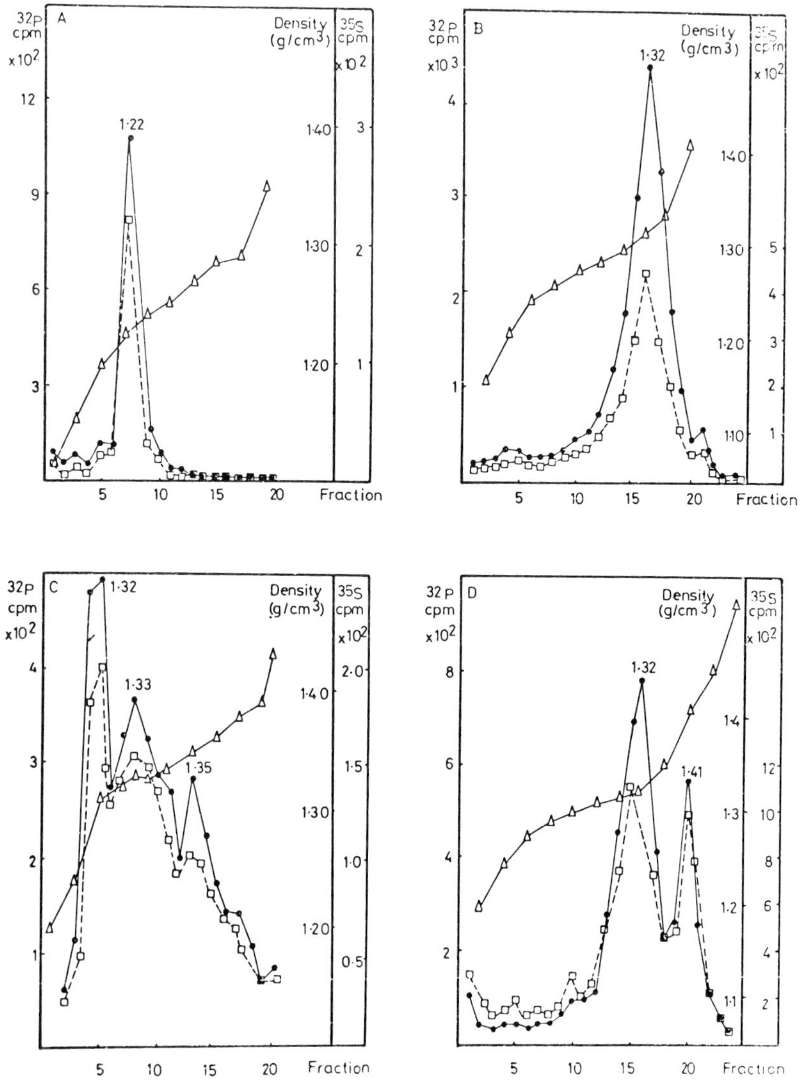

Figure 8. Effect of magnesium ions on the buoyant density of ribosomes in metrizamide. Ribosomes fixed with HCHO were centrifuged in gradients of 48% (w/v) metrizamide containing 10 mM Hepes-NaOH (pH 7.0) and either no Mg^{2+} (A); 3 mM Mg^{2+} (B); 10 mM Mg^{2+} (C) or 40 mM Mg^{2+} (D). The gradients were centrifuged in an MSE 10 x 10 ml titanium rotor at 105 000xg for 40 h at 5°C and the distribution of ^{35}S (■——■) and ^{32}P (●——●) in the gradients were analysed. With permission from ref. 6.

ribosomes was found to be 1.333 g/ml ± 0.002 (19). Thus, it appears that Nycodenz does not affect the composition of structure of the *E. coli* ribosomes *(Figures 9A* and *B)*. The 30S and 50S *E. coli* ribosomal subunits, purified through sucrose gradients and centrifuged either separately, or mixed together on preformed 20–70% (w/v) Nycodenz gradients at 160 000xg in an MSE 6x14 ml swing-out rotor at 5°C, banded at 1.17 g/ml *(Figure 9C)*.

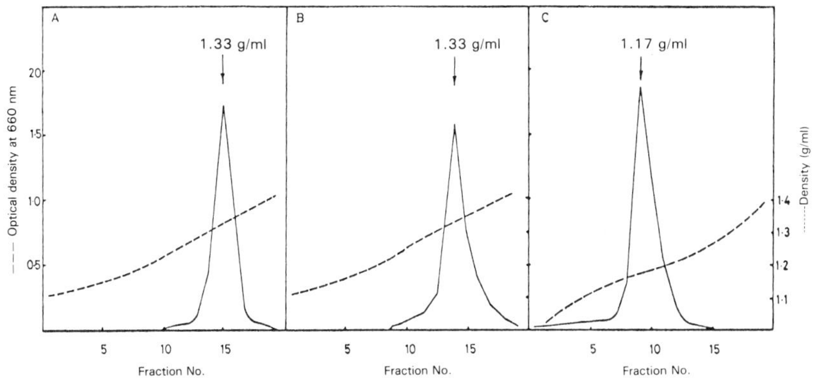

Figure 9. The density profiles of *E. coli* ribosomes in Nycodenz gradients. (A) *E. coli* 70S ribosomes, fixed prior centrifugation in 7% formaldehyde. (B) Unfixed *E. coli* 70S ribosomes. In (A) and (B), 5 ml gradients were made up of 4 ml 47% (w/v) Nycodenz, underlayered with 1 ml 60% (w/v) Nycodenz, and centrifuged 44 h at 75 000xg in an MSE 10 x 10 ml aluminium rotor at 5°C (19). (C) Unfixed *E. coli* ribosomal subunits, centrifuged on 20-70% (w/v) preformed Nycodenz gradients in the absence of Mg^{2+} at 160 000xg for 44 h at 5°C in an MSE 6 x 14 ml swing-out rotor (19).

3. SEPARATION OF NUCLEAR HETEROGENEOUS RIBONUCLEOPROTEINS

3.1. Introduction

The heterogeneous ribonucleoprotein particles (hnRNP) found in the nucleus are key structures because of their functional role in the post-transcriptional control of the genomic expression of cells. They include the high molecular weight nuclear RNA of the non-ribosomal primary transcripts, and the pre-messenger RNA products. Their complex structure is composed of single stranded RNA, RNA loops, small nuclear RNA (20), oligo(A), oligo(U) and poly(A) stretches. Numerous proteins are associated with these species of RNA, and we can ascribe to them putative structural roles as well as functional roles such as nuclease, ligase, methylase, maturase and other enzymatic activities. In fact, the hnRNP are now regarded as the nuclear fibrils (21) being closely associated with chromatin and the nuclear skeleton (22-24). After extraction from the nucleus, the hnRNP particles can be separated on the basis of size by centrifugation in sucrose gradients, the hnRNP sedimenting in an heterogeneous manner, ranging from 30S to more than 200S (25). It has been shown that, in separations on the basis of density in metrizamide gradients, the hnRNP gave rise to two peaks, one major peak at a density of 1.30 – 1.31 g/ml, and a smaller one at density 1.18 g/ml (26-28). These earlier studies have been extended, and the comparative migration of the hnRNP in metrizamide and Nycodenz is now analysed in this study.

3.2. Methods and Experimental Procedures

3.2.1. *Conditions of Culture and Labelling of Cells*

The Cl1D cells were grown and collected as indicated in Section 2.2. Labelling of RNA was performed in presence of low doses of actinomycin D. The use of actinomycin D, while it does not interfere with the hnRNP synthesis, blocks the ribosomal RNA synthesis and thus essentially eliminates isotopically labelled cytoplasmic polysomes and nucleolar ribosomes which might otherwise migrate in the same regions of the gradients; the actinomycin D provides an internal control in the experimental procedure. In some of the experiments, RNA was labelled with ^{3}H-adenosine (Amersham International plc) instead of ^{14}C-uridine, for 60 min or less, with 2.5 μCi/ml (92.5 kBq/ml) of medium. In all the other experiments the procedure for labelling RNA and protein was as in Section 2.2.

3.2.2. *Method for the Isolation of Purified Nuclei*

Resuspend the packed cells in 20 volumes of 0.25 M sucrose, 20 mM triethanolamine-HCl (pH 7.5), 3 mM magnesium acetate and 20 mM ammonium chloride. Add the nonionic detergent Nonidet P40 to a final concentration of 0.5%, and lyse the cells using 20 strokes in a glass-Teflon homogeniser. Pellet the crude nuclei by centrifugation at 5000xg for 10 min at 5°C and resuspend the nuclei in the same buffered solution. Add the detergent Triton X-100 to a final concentration of 1%, and homogenise the nuclei gently again. During all the operations, check the appearance of the nuclei by phase-contrast microscopy. When most of the cytoplasmic fragments have been removed, pellet the nuclei through a discontinuous sucrose gradient consisting of 55% (w/v) sucrose overlayered with 45% (w/v) sucrose containing 1% Triton X-100. Centrifuge the gradients at 30 000xg for 10 min, in a Beckman SW 27 or equivalent rotor. The pelleted nuclei are either used immediately, or stored at −70°C.

3.2.3. *Extraction and Centrifugation of hnRNP Particles*

Suspend the pellet of nuclei in 2.0 ml of 20 mM triethanolamine-HCl (pH 7.5), 100 mM NaCl, 25 mM KCl, and 2.5 mM magnesium acetate, in presence of ribonuclease inhibitors (cytoplasmic ribonuclease inhibitor, or 0.1% diethylpyrocarbonate). Sonicate the suspension of nuclei (MSE sonicator, 150 W), for 15 sec, in ice, at maximum power. As judged by phase-contrast microscopy, virtually all of the nuclei seem to be broken using these conditions. Centrifuge the sonicate twice at 30 000xg for 5 min in a Sorvall centrifuge. Discard the pellets, and immediately layer the supernatant either onto a 10−25% (w/v) sucrose gradient for rate-zonal separations giving the usual 30−200S heterogeneous pattern, or onto the iodinated solutions for isopycnic fractionations. The conditions of centrifugation and the handling of the gradients were as described in Section 2.2.4. Self-forming gradients of metrizamide, 41% or 45% (w/v) or of Nycodenz, 50% or 35−50% (w/v), as indicated, were used for all experiments.

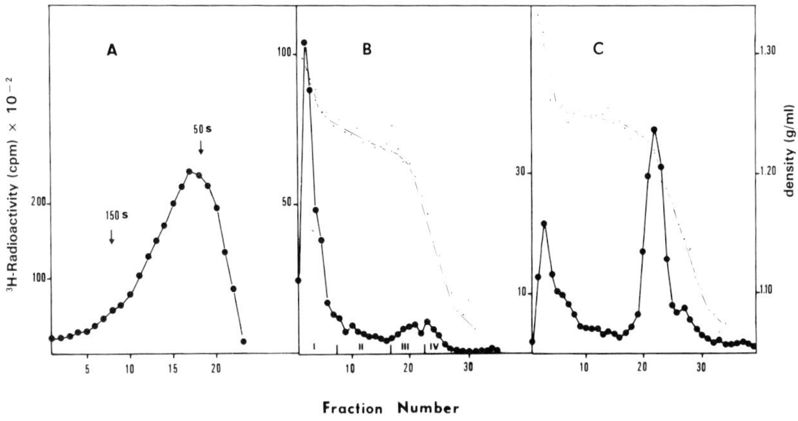

Figure 10. The separation of hnRNP particles on 10-25% (w/v) sucrose gradients and on 40% (w/v) metrizamide gradients The hnRNP particles were labelled with ^{3}H-adenosine for 1 h, in presence of actinomycin D (0.05 μg/ml). The isolation of hnRNP was carried out in presence of ribonuclease inhibitor (Searle). A. Sedimentation pattern of hnRNP in 10-25% (w/v) sucrose gradient. B. Density profile of hnRNP in 40% (w/v) metrizamide. C. Metrizamide gradient density profile of residual hnRNP isolated by a second extraction and washing of sonicated nuclei (28).

3.3. **Isopycnic Banding of hnRNP Particles**

3.3.1. *Separation in 41% (w/v) Metrizamide, and Size of the hnRNP Separated in the Different Fractions of the Density Gradient*

Heterogeneous nuclear RNA were labelled for 60 min with ^{3}H-adenosine, in the presence of actinomycin D. Nuclei and hnRNP were prepared as described in Sections 3.2.2. and 3.2.3. *Figure 10A* shows the sedimentation pattern of hnRNP in 10 – 25% sucrose gradient. The same material was layered concomitantly onto a 41% (w/v) metrizamide solution and centrifuged for 65 h in a Beckman SW 50.1 rotor or its equivalent. The density profile and distribution of material is shown in *Figure 10B* it consists of a major peak at about 1.30 g/ml, and a minor peak at 1.18 – 1.21 g/ml, with some material migrating between them. If, after the usual hnRNP extraction, the pellet of chromatin is re-extracted, one also gets two fractions *(Figure 10C)*, with, in this case, accumulation of RNP material in the 1.18 – 1.21 g/ml region of the gradient. To compare the size of the hnRNP in the different metrizamide fractions, to the size of the input hnRNP, the density gradient shown in *Figure 10B*, was fractionated into four parts, and the material of each part was directly layered onto 10 – 25% (w/v) sucrose gradients. The sedimentation patterns for each of the fractions are shown in *Figure 11*. The material of the major 1.30 g/ml peak gives a heterogeneous sedimentation profile similar to the original extract *(Figure 11A)* though with apparently an enrichment of high molecular weight RNP. The other fractions II, III, IV of the density gradient of *Figure 10B*, contain mostly low molecular weight material of 30S or less with, however, some apparently high molecular weight material also *(Figures 11B, C* and *D)*.

In a similar experiment, the length of poly(A) chains associated with the

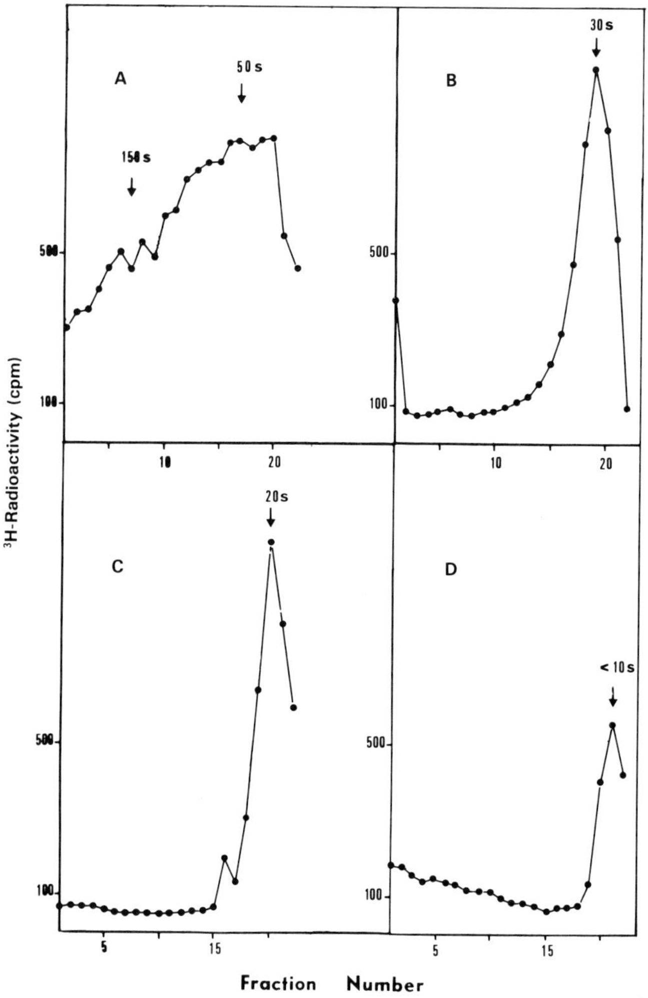

Figure 11. The sedimentation profiles in 10-25% (w/v) sucrose gradients of fractions of hnRNP particles previously separated in 40% (w/v) metrizamide. The metrizamide density gradient shown in *Figure 10B* was fractionated as indicated, into 4 parts corresponding to the density regions of the gradient, as follows: (A) Fraction I, denser than 1.25 g/ml; (B) Fraction II, 1.22 – 1.25 g/ml, (C) Fraction III, 1.18 – 1.22 g/ml, (D) Fraction IV lighter than 1.18 g/ml. The metrizamide fractions were diluted and put directly onto sucrose gradients, and centrifuged in a SW 27 rotor, at 4°C, 26 000 rev/min (98 000xg) for 3.5 h.

1.30 g/ml and the 1.18 – 1.20 g/ml hnRNP fractions were compared *(Figure 12)*. It is shown *(Figure 12A)* that the 1.30 g/ml hnRNP fraction, which corresponds to the high molecular weight particles, contains predominantly the large (150 – 200 nucleotides long) poly(A) chains, (*Figure 12A* peak I), while the 1.18 – 1.20 g/ml hnRNP fraction contains approximately equal amounts of poly(A) and oligo(A) chains (*Figure 12B* peaks I and II). The hnRNP material between 1.25 – 1.29 g/ml in density, gives a distribution pattern bet-

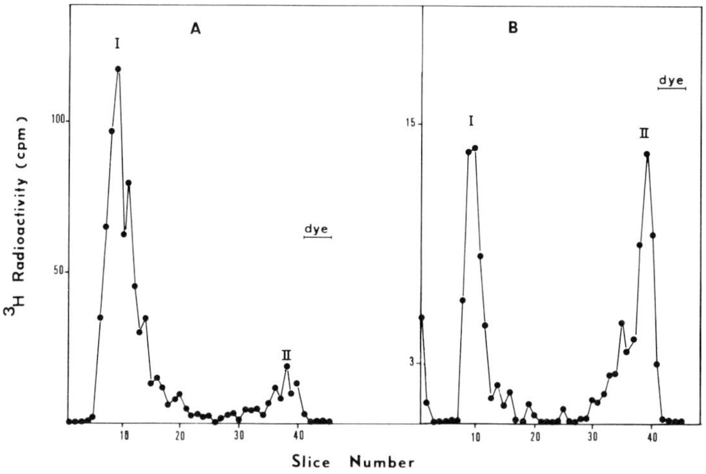

Figure 12. Electrophoretic migration of poly(A) and oligo(A) chains, purified from hnRNP fractions separated in 40% (w/v) metrizamide. Poly(A) and oligo(A) chains from: (A) hnRNP banding at density 1.28 – 1.31 g/ml, (B) hnRNP banding at density 1.18 – 1.22 g/ml. RNA was labelled with ^{3}H-adenosine for 20 min, in presence of 0.05 µg/ml actinomycin D. Cell nuclei were purified, and hnRNP particles extracted and separated on metrizamide gradients. The hnRNA from the 1.28 – 1.31 g/ml, and 1.18 – 1.22 g/ml density regions, was purified, and then hydrolysed by digestion with ribonucleases A and T_1 in presence of 0.3 M NaCl. The hydrolysate was chromatographed on oligo(U)-Sepharose, and then the poly(A) and oligo(A) chains were eluted with 50% formamide. Gel electrophoresis was carried out essentially according to Peacock and Dingman as modified by Darnell *et al.* (31). The final concentration of acrylamide was 10% (w/v).

ween poly(A) and oligo(A) intermediate to those shown on panel A and B of *Figure 12* (not shown). These observations on the size and the poly(A) content of hnRNP, are described as examples, to indicate that the separation of these nuclear particles in metrizamide, might be a useful means to study further their structure and functions.

3.3.1. *Comparative Analysis of the Banding of the hnRNP Particles, in 45% (w/v) Metrizamide, and 50% or 35 – 50% (w/v) Nycodenz*

The hnRNA was labelled with ^{14}C-uridine, in the absence *(Figures 13A_1, B_1, C_1)* or presence *(Figures 13A_2, B_2, C_2)* of actinomycin D. As expected, actinomycin D has little effect on the radioactivity of the extracted hnRNP, each gradient being loaded approximately with a similar amount of material. This actinomycin D treatment is useful for checking the nature of the isolated RNP material, especially if metrizamide is used, since polysomes and hnRNP both band in a similar manner in this iodinated medium (compare the panels C of *Figures 2, 3* and *13).*

Unexpectedly, the distribution profile of hnRNP particles in Nycodenz gradients was quite different from that observed in metrizamide. This was noted both in 50% (w/v) Nycodenz *(Figures 1A_1, A_2)* and in 35 – 50% (w/v) Nycodenz *(Figures 13B_1, B_2).* In the Nycodenz gradients, three peaks can be distinguished. It was also observed that the radioactivity profile of proteins does not exactly correlate with the radioactivity profile of RNA *(Figure 13A* and *B).*

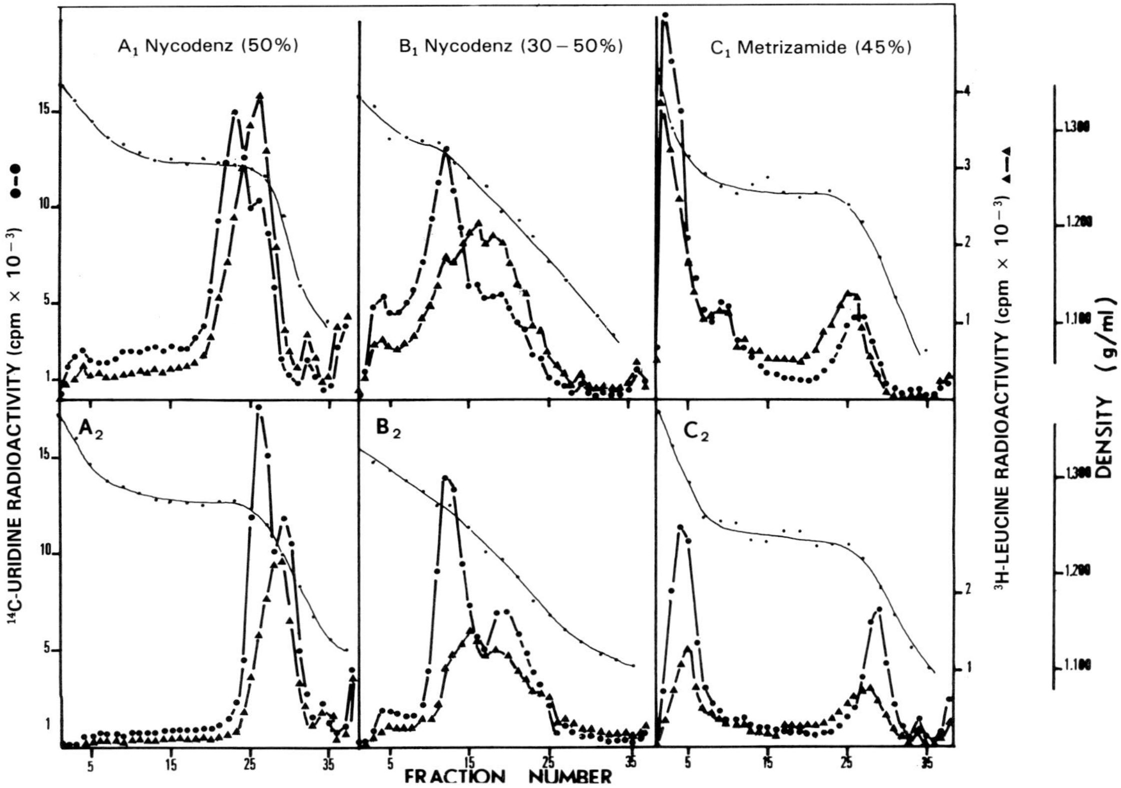

Figure 13. Comparative density profiles of hnRNP particles in Nycodenz and metrizamide gradients. (A) 50% (w/v) Nycodenz; (B) 35 – 50% (w/v) Nycodenz, in two superimposed layers; (C) 45% (w/v) metrizamide. The starting solutions, the buffer and centrifugation conditions, were as described in Section 2.2.4. and self-forming gradients were used. RNA and proteins were co-labelled for 4 h: (●——●) radioactivity of RNA (^{14}C-uridine), (▲——▲) radioactivity of proteins (^{3}H-leucine). The three upper panels ($A_1B_1C_1$) correspond to hnRNP labelled in the absence of actinomycin D. The three lower panels ($A_2B_2C_2$) correspond to hnRNP labelled in the presence of 0.05 μg/ml actinomycin D.

3.4. Advantages and Disadvantages of Using Iodinated Gradient Media for the Separation of hnRNP Particles

3.4.1. *The Distribution Profiles of the hnRNP Particles Separated in Metrizamide and Nycodenz Gradients*

As shown previously (26,28), and as illustrated in the experiments shown in *Figures 10B* and *13C*, the hnRNP particles migrate mainly as a major peak at a density of 1.30 – 1.31 g/ml in metrizamide gradients. This material, as judged by its heterogeneity and high molecular weight, remains unchanged after centrifugation for 65 h *(Figures 10A* and *11A)*. A minor part of the hnRNP material bands as a peak in the density region 1.18 – 1.20 g/ml, and is spread more or less continuously throughout the central part of the gradient. Although the low molecular weight material may contain possible degradation products of the larger molecules, dissociated RNA and proteins, its complete significance (20) has not yet been fully investigated. The preliminary observations reported here on the presence of residual apparently high molecular weight material, and on the distribution of the poly(A) and oligo(A) stretches, also need to be clarified by further studies.

The hnRNP particles, followed by thc radioactivity of their RNA banded in the Nycodenz gradients as three peaks *(Figures 13A* and *B)*: a minor peak at density of 1.30 – 1.31 g/ml, a major peak at density of 1.26 g/ml (1.25 – 1.27 g/ml), and a third peak at density of 1.22 g/ml (1.21 – 1.25 g/ml). These three fractions appear to be better separated in the near linear 35 – 50% (w/v) Nycodenz gradient *(Figure 13B)*. The sigmoid density gradient of *Figure 13A* might be useful for separating the peaks banding at 1.26 and 1.22 g/ml, by slightly decreasing the density of the plateau of the gradient. However, in this case, the 1.26 g/ml peak will become contaminated by the minor 1.30 g/ml peak. Here too, a full understanding of the nature of the hnRNP particles in the Nycodenz fractions (their size, poly(A) content, and a search for defined messenger RNA sequences) will require further study.

3.4.2. *Interactions Between the hnRNP Proteins and Iodinated Media*

As discussed earlier, it is known that metrizamide interacts weakly with proteins (14-16). On the other hand, the hnRNP particles should be viewed as fragile structures. These functional complexes are tightly integrated with other components of nuclei *in situ* and when they are extracted, these complexes are known to be sensitive to nucleases, other enzymes, and ions. Many of the hnRNP proteins are loosely associated with the hnRNA and hence during centrifugation in metrizamide gradients a significant fraction of these loosely bound proteins may become detached from the hnRNP (29). So these points should be borne in mind when using this iodinated medium; it is necessary to assess whether such effects are important for each type of fractionation.

Nycodenz has been found to interact less with proteins than does metrizamide (30), and hence one would expect the labelled RNA and proteins of the hnRNP to coband in Nycodenz gradients. The panels A_1, A_2 and B_1, B_2

of *Figure 13* show that, apparently, this is not the case. There is no parallel overlapping of the radioactivity curves; most of the radioactivity of proteins seems to accumulate in the 1.20 – 1.26 g/ml region of the Nycodenz gradients. This unexpected observation could be explained in terms of the hnRNP proteins dissociating from the RNA in the Nycodenz medium. Other hypotheses are also possible, and further studies of this phenomenon are necessary.

3.4.3. *An Assessment of the Usefulness of Iodinated Density Gradient Media for the Study of the hnRNP Particles*

Few methods have been developed to sub-fractionate hnRNP particles into structures related to particular functional states such as, primary transcripts, intermediary maturation products, potential pre-messenger particles and effective pre-messenger RNP. The reason for the interest in iodinated media is that they fractionate native hnRNP into several fractions banding at different densities, and the fractions obtained can be recovered and analysed further. The hnRNP banding pattern in Nycodenz gradients raises interesting questions. If the radioactivity profile of proteins was not artifactual, one explanation might be that the proteins associated with a sub-fraction of the hnRNP particles (namely the 1.20 g/ml peak, on *Figure 13B$_1$)* could have a higher turnover rate. Incidentally, it has been noted that, in the presence of actinomycin D, which affects protein synthesis, there is somewhat less radioactivity in the 1.20 g/ml peak *(Figure 13B$_2$)*. So iodinated media provide a new method for studying hnRNP particles. Further investigations will allow a fuller evaluation as to the value of metrizamide and Nycodenz gradients for separating functional entities of distinct hnRNP structures.

4. REFERENCES

1. Martin,R. and Ames,B. (1961) *J. Biol. Chem.,* **236**, 1372.
2. Perry,R.P. (1963) *Exp. Cell. Res.,* **29**, 400.
3. Osborn-Rivet,L. and Houssais,J.F. (1974) *Eur. J. Biochem.,* **48**, 427.
4. Buckingham,M.E. and Gros,F. (1975) *FEBS Lett.,* **53**, 355.
5. Buckingham,M.E. and Gros,F. (1976) in *Biological Separations in Iodinated Density Gradient Media* (Rickwood,D. ed.) IRL Press Ltd., Oxford and Washington, p. 71.
6. Rickwood,D. and Jones,C. (1981) *Biochim. Biophys. Acta,* **654**, 26.
7. Houssais,J.F., unpublished results.
8. Birnie,G.D. and Rickwood,D. (1976) in *Biological Separations in Iodinated Density Gradient Media* (Rickwood,D. ed.) IRL Press Ltd., Oxford and Washington, p. 193.
9. Rickwood,D. and Birnie,G.D. (1975) *FEBS Lett.,* **50**, 102.
10. Rickwood,D., Ford,T.C. and Graham,J. (1982) *Anal. Biochem.,* **123**, 33.
11. Henshaw,E.C. (1968) *J. Mol. Biol.,* **36**, 401.
12. Perry,R.P. and Kelley,D.E. (1968) *J. Mol. Biol.,* **35**, 37.
13. Dissous,C., Lempereur,C., Verwaerde,C. and Krembel,J. (1976) *Eur. J. Biochem.,* **64**, 361.
14. Hauge,J.G. (1971) *FEBS Lett.,* **17**, 168.
15. Infante,A.A. and Baierlein,R. (1971) *Proc. Natl. Acad. Sci. (USA)* **68**, 1780.
16. Rickwood,D., Hell,A., Birnie,G.D. and Gilhuus-Moe,C.C. (1974) *Biochim. Biophys. Acta,* **342**, 367.
17. Skerret,R.J. (1975) *Biochim. Biophys. Acta,* **385**, 28.
18. Hutterman,A. and Wendlberger-Schieweg (1976) *Biochim. Biophys. Acta,* **453**, 176.
19. Ford,T. (1983) M.Sc.Thesis, University of Essex, England.
20. Prüsse,A., Louis,C., Alonso,A. and Sekeris,C.E. (1982) *Eur. J. Biochem.,* **128**, 169.
21. Monneron,A. and Bernhard,W. (1969) *J. Ultrastruct. Res.,* **27**, 266.

22. Bachellerie,J.P., Puvion,E. and Zalta,J.P. (1975) *Eur. J. Biochem.,* **58**, 327.
23. Faiferman,I. and Pogo,A.O. (1975) *Biochemistry,* **14**, 3808.
24. Devilliers,G., Stevenin,J. and Jacob,M. (1977) *Biol. Cell.,* **28**, 215.
25. Jacob,M. and Stevenin,J. (1972) *Eur. J. Biochem.,* **29**, 480.
26. Houssais,J.F. (1975) *FEBS Lett.,* **56**, 341.
27. Houssais,J.F. (1976) in *Biological Separations in Iodinated Density Gradient Media* (Rickwood,D. ed.) IRL Press Ltd., Oxford and Washington, p. 81.
28. Houssais,J.F. (1977) *Mol. Biol. Rep.,* **3**, 251.
29. Gattoni,R., Stevenin,J. and Jacob,M. (1977) *Nucleic Acids Res.,* **4**, 3931.
30. Ford,T.C., Rickwood,D. and Graham,J. (1983) *Anal. Biochem.*, **28**, 293.
31. Darnell,J.E.,Jr., Wall,R. and Tushinski,R.J. (1971) *Proc. Natl. Acad. Sci. (USA),* **68**, 1321.
32. Noll,H. (1967) *Nature* **215**, 360.
33. Godson,G.N. and Sinsheimer,R.L. (1967) *Biochim. Biophys. Acta,* **149**, 489.
34. Bernabeu,C. and Lake,J.A. (1982) *Proc. Natl. Acad. Sci. (USA),* **79**, 3111.
35. Noll,H. (1969) in *Techniques in Protein Synthesis* (Campbell,P.N. and Sargent,J.R. eds) Academic Press Inc., New York and London. Vol **2**, p. 101.
36. Gielkens,A.L.J., Berns,T.J.M. and Bloemendal,H. (1971) *Eur. J. Biochem.,* **22**, 478.
37. Falvey,A.K. and Staehelin,T. (1970) *J. Mol. Biol.,* **53**, 1.
38. Spohr,G., Granboulan,N., Morel,C. and Scherrer,K. (1970) *Eur. J. Biochem.,* **17**, 296.

CHAPTER 4

Preparation and Fractionation of Nuclei, Nucleoli and Deoxyribonucleoproteins

D. RICKWOOD and T.C. FORD

1. INTRODUCTION

DNA is a long fibrous molecule and in nature it is always found associated with proteins and sometimes other basic molecules. The proteins found associated with the DNA are involved not only in its replication and transcription but also in the packaging of the DNA into compact structures. An idea of the magnitude of the problems in packaging the DNA can be gauged by considering that eukaryotic cells can contain a metre or more of DNA that must be packaged into a nucleus usually only about 3 μm in diameter. In eukaryotes the DNA is combined with a specific group of basic proteins, the histones, to form bead-like structures called nucleosomes. In turn, nucleosomes are organised into higher order structures (1). In interphase nuclei, the chromatin appears as an amorphous mass within an intranuclear proteinaceous scaffold (2). The most prominent feature of eukaryotic nuclei are the nucleoli which are compact structures responsible for the synthesis of ribosomes and associated with the genes coding for the ribosomal RNA. The only time that chromatin of eukaryotic cells forms distinct structures is during mitosis and meiosis when the chromatin forms a number of distinct condensed chromosomes; the number of chromosomes varies depending on the type of organism.

Prokaryotes have a single chromosome which is attached to the cell membrane. However, less is known as to how the DNA is folded into compact structures in prokaryotes although some basic proteins have been found associated with the DNA (3). In the case of mitochondria and chloroplasts, these organelles of eukaryotic cells contain much less DNA than bacteria, but it appears that the DNA is packaged into nucleoids (4) in a similar fashion to bacterial DNA. Usually the DNA of organelles is in the form of a single chromosome.

However, in no case is the DNA continually in the form of a compact structure. The DNA must always unfold in order to replicate and to allow the synthesis of RNA from the DNA template. Both of these processes involve the association of additional proteins with the DNA and, as a result of the different compositions of these complexes, there is the possibility of fractionating them from the bulk of the DNA.

A number of standard methods have been devised for the isolation of nuclei, nucleoli and chromatin as well as for some types of nucleoid from bacteria and cell organelles. The nuclei of some types of cells can be purified isopycnically by sedimenting them through dense sucrose (5), although this is not suitable for all tissues. It is not possible to fractionate different types of nuclei by isopycnic centrifugation in sucrose gradients. Nucleoli can be freed from their surrounding nucleoplasm by disrupting purified nuclei and the nucleoli can be purified from the nuclear debris by differential centrifugation. Chromosomes have also been purified by differential and rate-zonal centrifugation but, in this case, it is important to use the correct medium to maintain the morphology of the chromosomes (6). Chromatin can be prepared by a wide variety of methods (7) although most involve differential extraction of the soluble material of nuclei with buffer solutions to remove the ribonucleoproteins; these procedures can result in significant losses of chromatin. Isopycnic fractionation of chromatin has been carried out using CsCl gradients. However, the high ionic strength of CsCl gradients means that the chromatin must first be fixed using cross-linking agents such as formaldehyde or glutaraldehyde to prevent the dissociation of the proteins from the DNA; such fixation makes subsequent analysis of chromatin very difficult.

Ionic iodinated compounds (e.g. sodium metrizoate) have been tried for isopycnic fractionations of nuclei, chromosomes and chromatin. However, although the ionic strength of these solutions is significantly less than those of CsCl, they are still able to disrupt nucleoprotein structures (8). With the introduction of metrizamide in 1973 it became feasible to fractionate nuclei, nucleoli, chromosomes and chromatin isopycnically in ionic environments that optimise the retention of the native state of the sample. It has been found that neither metrizamide nor Nycodenz[1] affects the morphology of nuclei (31). In a number of cases the fractionations obtainable in metrizamide gradients are markedly better than if other gradient media are used. This chapter describes in detail the applications of gradients of metrizamide and Nycodenz for the isolation and fractionation of nuclei and deoxyribonucleoprotein structures. Because metrizamide became available several years before Nycodenz was introduced, most of the methods described involve the use of metrizamide gradients. However, it is likely that similar separations can be carried out using Nycodenz gradients and indeed in some cases there are significant advantages in using Nycodenz instead of metrizamide because it interferes less with chemical assays (see section 5.2.2 of Chapter 1). Where possible comparative data for both metrizamide and Nycodenz are given.

2. PREPARATION AND FRACTIONATION OF NUCLEI

2.1. Purification of Nuclei

The standard method for the purification of nuclei involving centrifugation through dense sucrose solutions was originally devised for liver nuclei (5). This method works well with cells that are easy to break by homogenisation, usually as a result of the cells having a high ratio of cytoplasm to nucleus. The method also depends on the assumption that the nuclei are the densest particles of the cell and, where this is so, the method usually works well but in other cases the nuclei can be extensively contaminated with other non-nuclear material.

A typical problematical tissue is epidermis. The sucrose method cannot be used for this tissue because the nuclei tend to be contaminated with membrane

[1]Nycodenz is a registered trademark of Nyegaard & Co., Oslo, Norway.

Table 1. Purification of Nuclei from Epidermis.

1. Place epidermal strips, excised immediately after sacrifice (9), in water at 56°C for 30 sec followed by incubation in iced-water for 15 sec to separate the epidermis from the dermis. Separate the two using fine forceps.
2. Mince the epidermal tissue in 3% citric acid, 0.1 mM phenylmethylsulphonyl fluoride (PMSF), and then homogenise the suspension using 10 strokes of a Dounce homogeniser. Centrifuge the homogenate at 3000xg for 15 min at 5°C.
3. Resuspend the pelleted nuclei in the homogenisation buffer and filter the suspension through 3 successive Nylon filters with pore sizes of 105 μm, 30 μm and 20 μm respectively.
4. Pellet the nuclei by centrifugation at 3000xg for 15 min, resuspend in homogenisation buffer and check the purity of the nuclei, if more than 5% of the nuclei have cytoplasmic adhesions then repeat Step 2.
5. Load the nuclei (3 x 10^7 nuclei) onto a 6 ml gradient of 40–60% (w/v) metrizamide (Nyegaard & Co., Oslo, Norway) gradient containing 3% citric acid and centrifuge the gradients at 49 000xg for 60 min. The nuclei band at 1.30 g/ml.
6. Recover the nuclei from the gradient fractions by centrifugation at 5000xg for 10 min after dilution with 2 volumes of 3% citric acid.

material. Instead a method has been devised which combines the use of citric acid to remove cytoplasmic contamination and metrizamide gradient centrifugation to isolate the purified nuclei which band at much higher densities than membranes (9). *Table 1* gives the full protocol for the purification of epidermal nuclei. The overall yield of nuclei using this method is about 40%.

Examples of tissues containing dense particles which are potential contaminants of sucrose-prepared nuclei are the tissues of *Xenopus laevis* and other amphibians that contain dense melanosomes. An alternative to the sucrose procedure is to purify the nuclei by banding them in metrizamide gradients since melanosomes band at a higher density than nuclei (10). *Table 2* describes the procedure that should be used. Adult liver cells of *X. laevis* contain more melanosomes than embryonic cells and so it is necessary to include two separate metrizamide purification steps (Steps 4 and 7 of *Table 2*). In these metrizamide gradients the melanosomes pellet while the purified embryo and liver nuclei band at 1.288 g/ml and 1.284 g/ml, respectively; any nuclei contaminated with attached melanosomes band at higher densities. The average yields of embryo and liver nuclei are $48 \pm 8\%$ and $68 \pm 7\%$, respectively. *Figure 1* illustrates the much greater purity of embryonic nuclei purified in metrizamide gradients as compared with those purified by centrifugation through 2.4 M sucrose. This method is likely to be applicable to all types of amphibian tissues in which melanosomes are present and indeed it should be applicable to the purification of all types of nuclei from cells containing dense particles likely to contaminate nuclei centrifuged through dense sucrose solutions.

2.2. Fractionation of Nuclei

Most eukaryotic tissues consist of a number of different cell types. For example, liver consists mainly of parenchymal cells but as many as 30% of the cells are non-

Table 2. Purification of Nuclei from *X. laevis*.

1. Homogenise embryos (500/100 ml) or minced livers (1 g/10 ml) in 0.25 M sucrose in buffer A which contains 3 mM $CaCl_2$, 3 mM Tris-HCl (pH 8.0), 50 mM $NaHSO_3$, 1 mM PMSF. Homogenise the cell suspensions in a motor-driven Teflon-glass Potter-Elvejhem homogeniser (clearance 0.15 – 0.23 mm) at 1200 rev/min using one stroke for embryos and six strokes for liver.
2. Filter the homogenate through a Nitex filter with 100 μm pores and centrifuge the filtrate at 1000xg for 10 min at 5°C.
3. Using the homogeniser driven at 600 rev/min, resuspend the pellet of crude nuclei in the homogenising medium containing 1% Triton X-100 and pellet the nuclei by centrifugation at 1000xg for 10 min at 5°C.
4. Resuspend the Triton X-100 washed nuclei in 2.0 ml of homogenisation buffer and add 8.0 ml of 72.5% (w/v) metrizamide (Nyegaard & Co., Oslo, Norway) dissolved in buffer A to give a final concentration of 58% (w/v). Disperse the aggregates by gentle hand homogenisation and centrifuge the solution at 10 000xg for 20 min at 5°C. The melanosomes sediment through the solution while the nuclei form a pellicle at the surface. Remove the pellicle of nuclei using a spatula.
5. Suspend the nuclei in 9 vols of 2.3 M sucrose in buffer A by homogenisation as in Step 3. Layer the suspension of nuclei over 5.0 ml of 2.4 M sucrose also in buffer A and centrifuge at 120 000xg for 60 min at 5°C in a swing-out rotor.
6. Wash the nuclear pellets, contaminated with some melanosomes, in homogenisation medium containing 1% Triton X-100 (Step 3).
7. Resuspend the washed, pelleted nuclei in homogenisation medium and layer the suspension (10^9 nuclei/35 ml gradient) onto a 40 – 60% (w/v) metrizamide gradient prepared in buffer A. Centrifuge the gradient at 120 000xg for 60 min at 5°C.
8. Unload the gradients by upward displacement with Maxidens (Nyegaard & Co.) and locate the fractions containing nuclei. Dilute the fractions by the addition of two volumes of buffer A and pellet the nuclei by centrifugation at 5000xg for 10 min at 5°C.

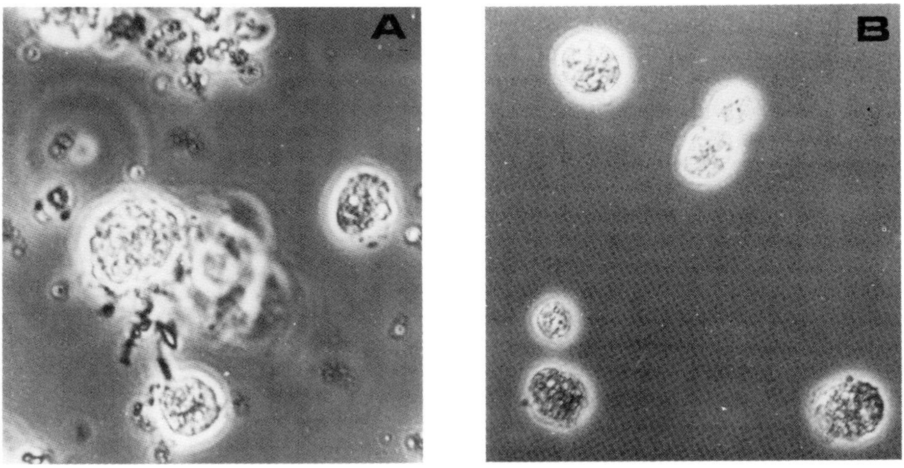

Figure 1. A comparison of embryonic *X. laevis* nuclei prepared by centrifugation through either (A) 2.4 M sucrose or (B) metrizamide. Data derived from ref. 10.

parenchymal cells consisting of several types. Frequently the different types of cells in a tissue have different metabolic activities and these differences are reflected not only in the transcriptional activity of the DNA but also in the relative amounts of protein, RNA and DNA in the nucleus, generally more active nuclei contain larger amounts of RNA and protein. These differences in the composition of nuclei introduce the possibility of separating the nuclei of these different types of cells. To obtain good separations of nuclei, it is important to optimise the density range of the gradient to ensure the maximum degree of resolution of the different types of nuclei present.

Brain nuclei have been separated on metrizamide gradients by both rate-zonal and isopycnic centrifugation (11). Both types of separation fractionate the nuclei from neuronal, astrocytic and oligodendroglial cells into different classes. It is found that the nuclei from neuronal cells which have the highest transcriptional activity band at the highest density (*Figure 2*). Nuclei from *X. laevis* liver also band over a wide density range and it is possible to separate the nuclei of non-parenchymal cells from those of parenchymal cells since the latter band at lower

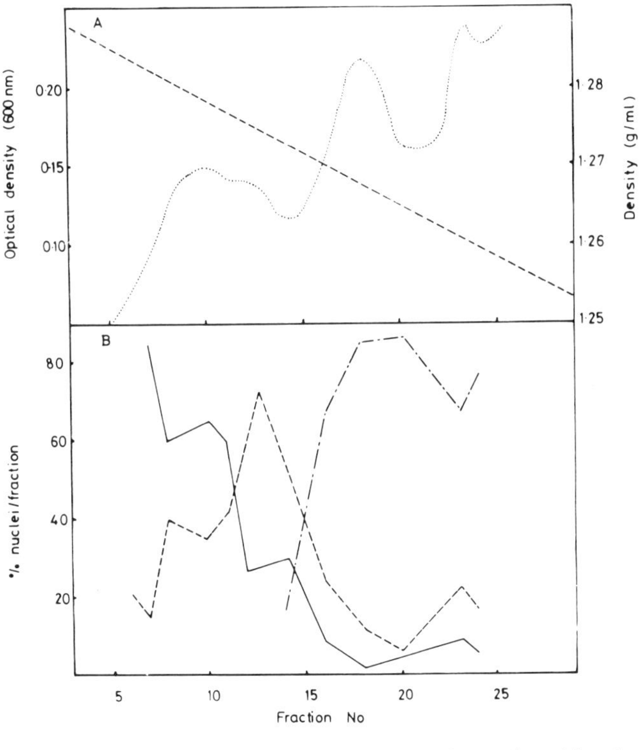

Figure 2. Fractionation of nuclei from rat cerebrum sedimented to isopycnic positions in a linear gradient of metrizamide containing 0.32 M sucrose. (A) Shows distribution of total nuclei throughout the gradient as measured by light scattering at 600 nm (· · · · ·) and the density profile of the gradient (— — —). (B) The percentages of the nuclei of different classes in each fraction are shown for neuronal nuclei (———), astrocytic nuclei (— —), oligodendroglial nuclei (—·—·). The peak in fractions 23 and 24 is mainly derived from fragmented nuclei. Data derived from ref. 11.

Table 3. Purification of Tetrahymena Nucleoli on Metrizamide Gradients (14).

1. Suspend purified nuclei in 0.34 M sucrose, 0.1 mM $MgCl_2^a$, 10 mM Tris-HCl (pH 7.4), and sonicate the suspension; usually two or three 20 sec periods of sonication with cooling between each is sufficient. Check the extent of nuclear disruption by light microscopy; if necessary sonicate further.
2. Pellet any large aggregates by centrifugation at 80xg for 5 min. Discard the pellet and sediment the nucleoli at 13 000xg for 20 min.
3. Resuspend the pellet in 30% (w/v) metrizamide (Nyegaard & Co., Oslo, Norway), 90 mM NaCl, 1.5 mM KCl, 0.1 mM $MgCl_2$, 10 mM Tris-HCl (pH 7.4), 0.1% Triton X-100 and mix it with 60% (w/v) metrizamide in the same buffer using a gradient mixer to form a linear gradient.
4. Centrifuge the gradient at 125 00xg for 1.5 h at 4°C.
5. Fractionate the gradients and locate the nucleoli by light-scattering measurements at 660 nm.
6. Dilute the fractions containing nucleoli with two volumes of 1.0 mM $MgCl_2$ and pellet the nucleoli by centrifugation at 13 000xg for 10 min.

[a]For some other cell types 3.3 mM $CaCl_2$ has been used instead of $MgCl_2$.

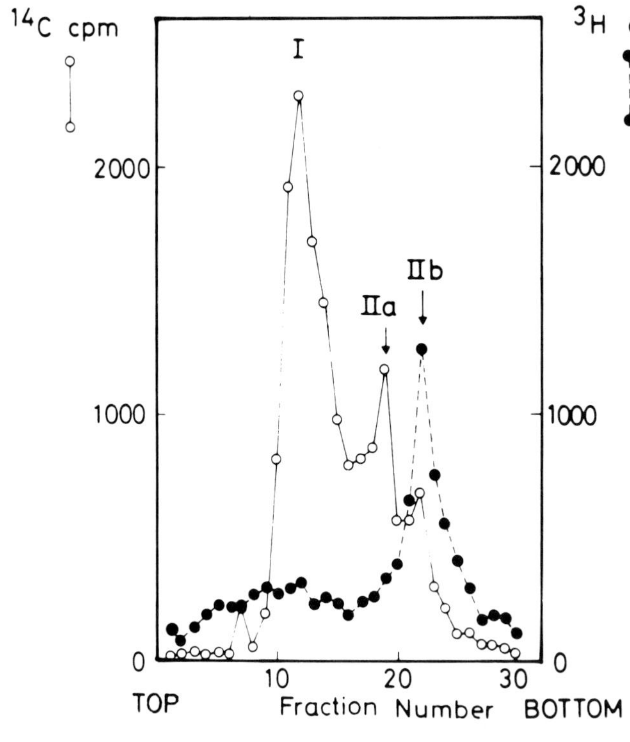

Figure 3. Metrizamide density gradient centrifugation pattern of a tetrahymena nuclear sonicate. Nuclei were isolated from tetrahymena cells in which the rDNA had been preferentially labelled with ^3H-thymidine (●——●) by the technique of starving and re-feeding (14); the bulk DNA was labelled with ^{14}C-thymidine (○——○). Isolated nuclei were sonicated and the sonicate was loaded onto a metrizamide gradient. The gradients were centrifuged at 80 000xg for 1.5 h at 4°C. An aliquot from each fraction was placed onto a glass-fibre disk (Whatman GF/C) and assayed for ^{14}C and ^3H radioactivity. Data from ref. 14 reproduced with permission of the authors and publishers.

densities (10). Nuclei of different developmental stages can also be fractionated. As an example, it has been shown that nuclei of rat spermatids can be fractionated in metrizamide gradients on the basis of the stage of development of the cells (12).

3. PURIFICATION OF NUCLEOLI

The original methods for preparing nucleoli involved destruction of the nuclear structure by ultrasonication or DNase digestion followed by pelleting of the liberated nucleoli by differential centrifugation. The inefficiency of purification procedures involving differential centrifugation have been discussed elsewhere (13) and usually it is necessary to compromise either on the purity of the nucleoli or the yield obtained. Isopycnic centrifugation in gradients of metrizamide or Nycodenz offers the possibility of obtaining a pure population of nucleoli minimally cross-contaminated with extra-nucleolar material. Such methods are particularly useful for types of tissue in which the nuclei are not amenable to simple differential centrifugation techniques.

An example of the type of procedure that can be used for the purification of

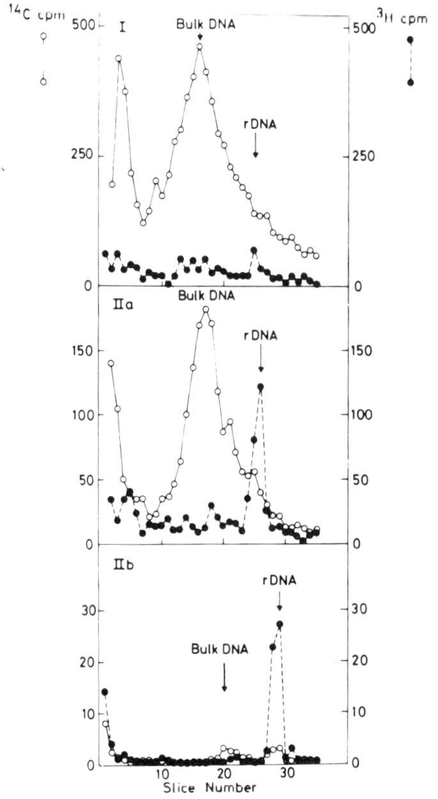

Figure 4. Agarose gel electrophoresis profile of DNA extracted from each peak of *Figure 3*. The fractions corresponding to Peaks I, IIa, and IIb of *Figure 3* were combined and extracted with phenol. DNA was precipitated from the aqueous phase and analyzed by electrophoresis on 0.6% agarose gels. Data from ref. 14 reproduced with permission of the authors and publishers.

nucleoli on metrizamide gradients is given in *Table 3*. The type of fractionation obtained in the case of Tetrahymena nucleoli is shown in *Figures 3* and *4*. Nucleoli have also been purified from other types of cell, notably from *X. laevis* oocytes (15) in which the nucleoli are highly amplified. However the presence of the other components of oocytes (e.g. melanosomes and yolk particles) precludes the use of the usual differential centrifugation techniques. In order to obtain purified nucleoli, Dounce homogenise the oocytes in 90 mM NaCl, 1.5 mM KCl, 5 mM $MgCl_2$, 5 mM 2-mercaptoethanol, 0.1 mM EDTA and 50 mM Tris-HCl (pH 7.4). Filter the homogenate through four layers of fine gauze and add an equal volume of 65% (w/v) metrizamide (Nyegaard & Co., Oslo, Norway) solution to give a final concentration of 32.5% (w/v). Place this metrizamide solution containing the nucleoli into the chamber of a simple gradient mixer and, using a 50% (w/v) metrizamide solution in the same buffer as the denser solution, form a linear gradient. Centrifuge the gradients at 120 000xg for 60 min at 5°C. Unload the gradient and determine the position of the nucleoli by light-scattering measurements at 650 nm. The nucleoli band at 1.24 g/ml, the extranucleolar material bands at 1.22 g/ml and unbroken follicle cells band at 1.21 g/ml. Nucleoli can be recovered from fractions by centrifugation after dilution with buffer. Using this method there is no significant contamination of the nucleoli after purification on a metrizamide gradient.

4. PURIFICATION OF METAPHASE CHROMOSOMES

Chromatin condenses to form distinct chromosomes during mitosis but usually

Table 4. Procedure for the Isopycnic Banding of Metaphase Chromosomes.

1. Soak 15 ml centrifuge tubes, plastic or glass (Corex grade), for 5 min in 1% dimethyldichlorosilane in carbon tetrachloride (Repelcote, Hopkin and Williams, Ltd.; Siliclad, Clay Adams).
2. Drain the tubes and bake them at 60° – 80°C for 24 h. Once prepared these tubes can be stored for use at a later date.
3. Wash the cells, arrested in metaphase by treatment with Colcemid (0.06 μg/ml for 3 h), with cold lysis buffer containing 1.0 M hexylene glycol, 0.5 mM $CaCl_2$, 0.1 mM Pipes-NaOH (pH 6.5) and pellet the cells by centrifugation at 600xg for 3 min at 4°C. Suspend the cells in cold lysis buffer and incubate the cell suspension at 37°C for 10 min. Lyse the cells either by passing them gently through a 22 gauge needle or by nitrogen cavitation. Pellet the metaphase chromosomes by centrifugation at 2000xg for 5 – 10 min.
4. Suspend the metaphase chromosomes in 1.0 ml of 60% (w/v) metrizamide (Nyegaard & Co., Oslo, Norway) containing 1.0 M hexylene glycol, 5 mM 2-naphthyl-6,8-disulphonic acid, 0.5 mM $CaCl_2$, 0.1 mM Pipes-NaOH (pH 6.8).
5. Place the chromosome suspension in a siliconised 15 ml centrifuge tube and overlayer it with a linear gradient of 25% – 60% (w/v) metrizamide containing 1.0 M hexylene glycol, 5 mM 2-naphthyl-6,8-disulphonic acid, 0.5 mM $CaCl_2$, 0.1 mM Pipes-NaOH (pH 6.8).
6. Centrifuge the gradients at 16 300xg for 30 min at 5°C. During centrifugation the chromosomes float upward to form a sharp band which is detectable with the naked eye.
7. Unload the gradients by upward displacement.
8. Recover the metaphase chromosomes by centrifugation after diluting the fractions with two volumes of gradient buffer.

only a small percentage of a growing cell population is in mitosis. In order to obtain a high percentage of cells in the mitotic stage it is necessary to use an inhibitor such as Colcemid or one of its derivatives which arrest the cells at metaphase stage of mitosis. Using this technique with higher eukaryotic cells one can obtain as many as 95% of the cells in metaphase (16). However this method is much less effective in obtaining metaphase arrest in lower eukaryotic cells.

In order to isolate intact metaphase chromosomes, the cells should be lysed as gently as possible; nitrogen cavitation usually gives the best results (17) although

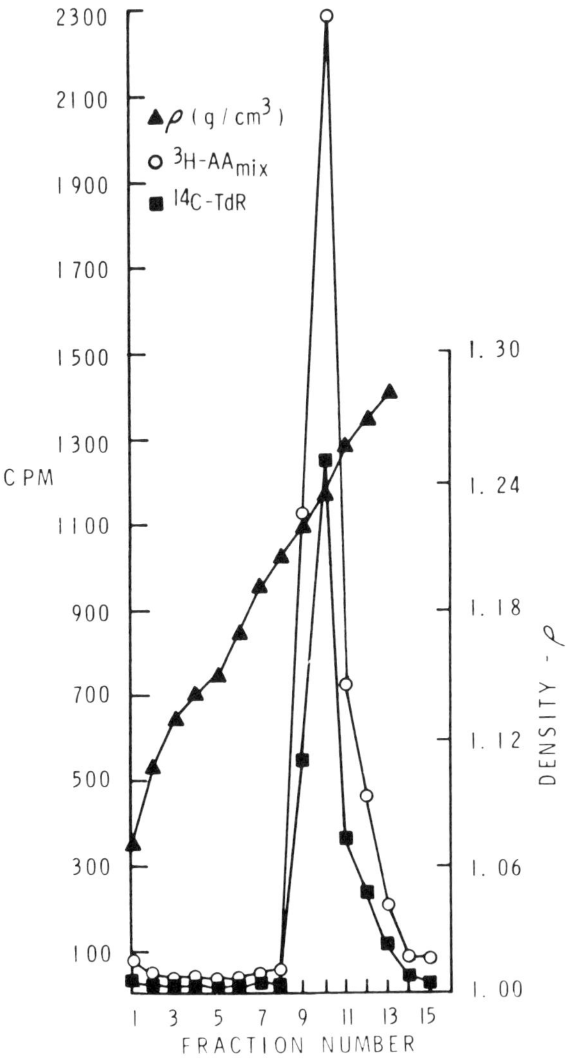

Figure 5. Isopycnic banding of metaphase chromosomes doubly-labelled with ^{14}C-thymidine (■——■) and ^{3}H-amino acids (○——○). Tubes containing 5 ml of a preformed 0.35 M to 0.75 M metrizamide-buffer gradient were prepared and centrifuged as described in *Table 4*. Data from ref. 8.

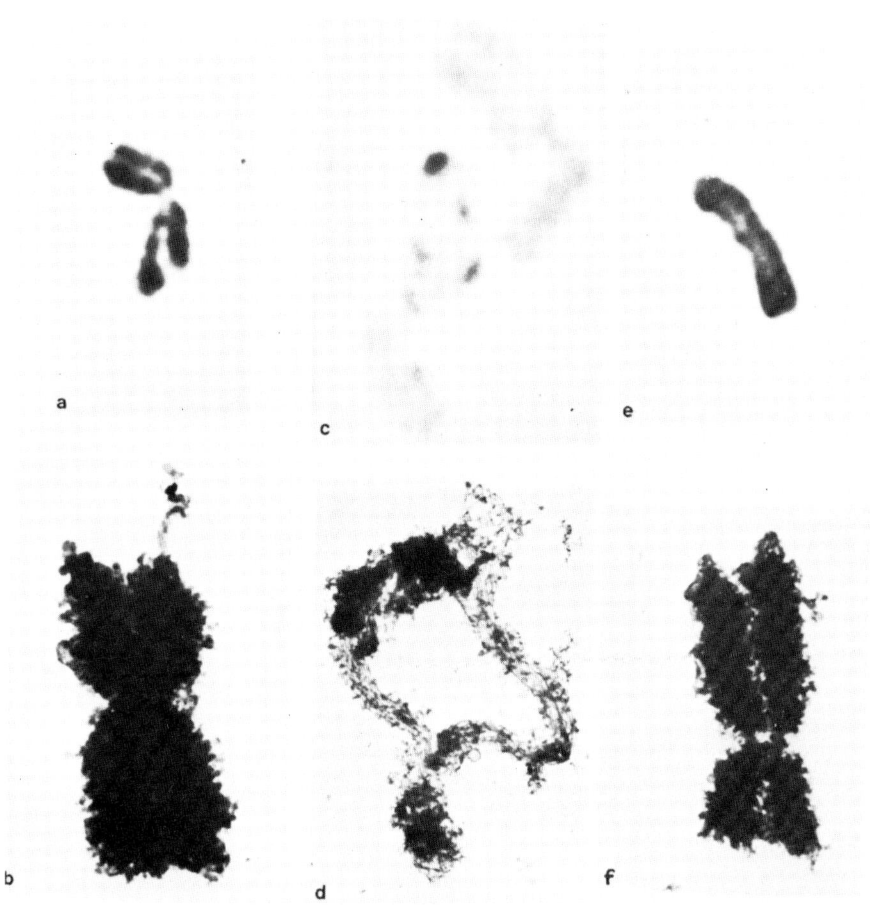

Figure 6. Effect of sodium diatrizoate and metrizamide on chromosome morphology. (a) Phase-contrast light micrographs of an isolated chromosome (control). Magnification x 1000. (b) Electron micrograph of an isolated chromosome (control). Magnification x 17 000. (c) Phase-contrast light micrograph of isolated chromosome exposed to sodium diatrizoate. Magnification x 1000. (d) Electron micrograph of isolated chromosome exposed to sodium diatrizoate. Magnification x 9900. (e) Phase-contrast light micrograph of isolated chromosome exposed to metrizamide. Magnification x 1000. (f) Electron micrograph of isolated chromosome exposed to metrizamide. Magnification x 9900. Data from ref. 8.

other methods of gentle cell lysis can also give reasonably intact preparations. The other extremely important requirement is to use a buffer which will stabilise the structure of the chromosomes once the cells are lysed (6). In order to isolate the chromosomes at neutral pH one should use a lysis buffer containing 1.0 M hexylene glycol (2-methyl-2,4-pentanediol), 0.5 mM $CaCl_2$, 0.1 mM Pipes-NaOH (pH 6.8). Once the cells are lysed the chromosomes can be banded in metrizamide gradients as described in *Table 4*. The flotation method given in *Table 4* overcomes two of the major difficulties of working with metaphase chromosomes namely the tendency of the chromosomes to aggregate with each other and to stick to walls of the centrifuge tube. It has been found that the best results are obtained using short centrifugation times.

Using this technique the chromosomes form a single sharp band in metrizamide gradients (*Figure 5*). Electron microscopy shows that, unlike ionic iodinated gradient media, metrizamide does not disrupt the structure of metaphase chromosomes (*Figure 6*). Chromosomes can also be banded in Nycodenz gradients since it also does not affect the morphology of metaphase chromosomes (31). It has been shown that acid-extraction of chromosomes modifies their buoyant densities (17) and this raises the possibility that resolution of different chromosomes might be achieved by treating the chromosomes using one of the differential banding techniques (e.g. quinacrine mustard) prior to centrifugation in a metrizamide gradient. As in the case of macromolecules (see Chapter 1) it is likely that chromosomes can be separated using density labelling techniques.

5. PURIFICATION AND FRACTIONATION OF CHROMATIN

5.1. General Features of Chromatin Banded in Metrizamide and Nycodenz Gradients

As shown in Section 4 of this chapter and in Chapter 3, nonionic iodinated gradient media do not appear to dissociate nucleoprotein complexes. In addition, not only does metrizamide not affect the higher order structures of metaphase chromosomes but also metrizamide does not disrupt the nucleosome structure of chromatin (*Figure 7*).

As expected from the differences in the buoyant densities of nucleic acids and proteins (see Chapter 2), the density of chromatin depends on its composition in terms of DNA, RNA and protein. However, the buoyant density of chromatin in metrizamide and Nycodenz gradients is also dependent on factors that modify the conformation of chromatin since changes in conformation can affect the hydration of the chromatin.

In the absence of ions, chromatin is readily soluble in solutions of metrizamide and Nycodenz and high molecular weight chromatin bands as a single component. When chromatin is banded in metrizamide gradients containing sub-millimolar amounts of ions the buoyant density of the chromatin depends on the amount of chromatin loaded onto the gradient. Increasing the amount of chromatin loaded onto a metrizamide gradient increases the buoyant density (18); a similar but smaller effect is also seen in the case of Nycodenz gradients (*Figure 8*). This effect can be explained in terms of aggregation of the chromatin as it forms a band in the gradient. The higher the concentration of the chromatin in the band the higher is the degree of aggregation, leading to a decrease in the hydration of the chromatin. It is not clear why this effect of loading is smaller in the case of Nycodenz gradients. This effect is seen not only in the case of chromatin but nucleoids from organelles (see Section 6) also show the same effect. However, no such effect is seen with ribonucleoproteins in either medium.

Increasing the concentration of NaCl in the gradient increases the buoyant density of chromatin reaching a maximum of 0.14 M NaCl; at higher concentrations

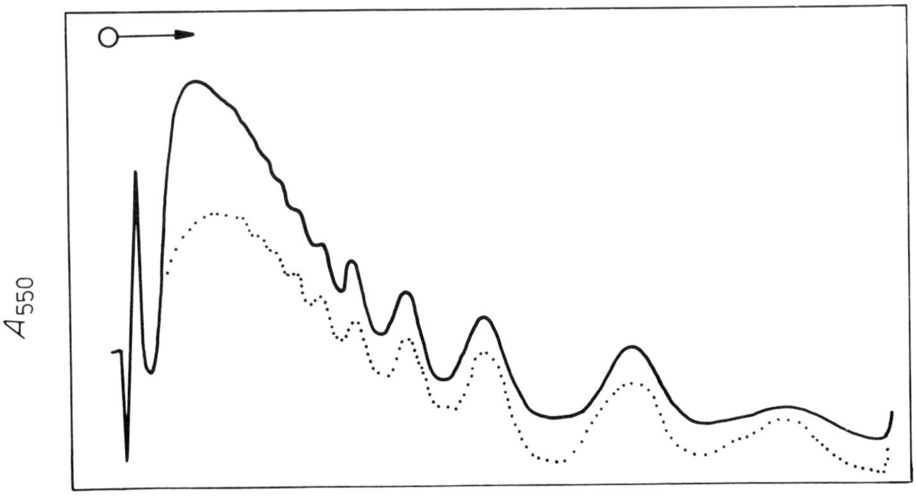

Figure 7. The effect of metrizamide on the structure of nucleosomes. Purified mouse-liver nuclei were suspended in 0.1 mM $CaCl_2$ and 2 mM Tris-HCl (pH 7.4) containing either 0.25 M sucrose or 40% (w/v) metrizamide for 18 h at 0°C. Each suspension was then digested with staphylococcal nuclease at 37°C for 10 min. The reaction was terminated by the addition of EDTA, and DNA was prepared separately from the reaction mixtures containing sucrose (———) or metrizamide (· · · · ·) prior to electrophoresis in 2.5% polyacrylamide gels (32). Data reproduced from ref. 33 with permission of the authors and publishers.

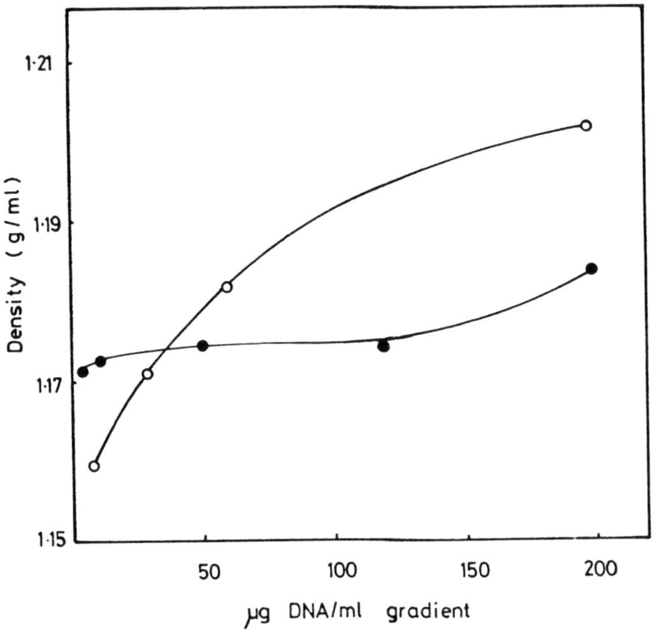

Figure 8. Effect of loading on the buoyant density of chromatin in metrizamide and Nycodenz gradients. Chromatin isolated from mouse-liver nuclei was solubilised (27) and loaded separately onto metrizamide and Nycodenz gradients and centrifuged as described elsewhere (27). The variation of buoyant density with loading in metrizamide (○——○) and Nycodenz (●——●) gradients is shown.

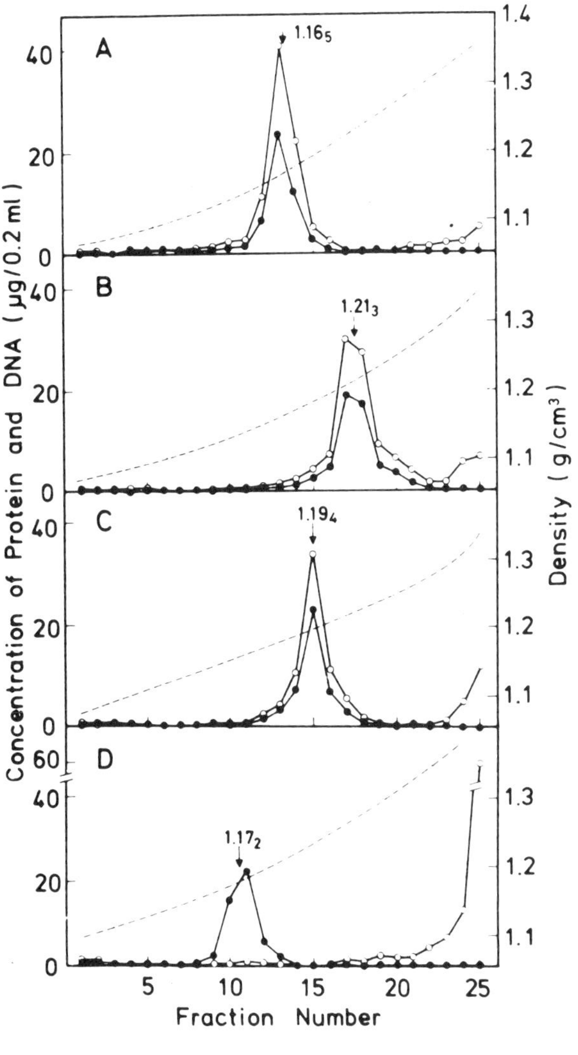

Figure 9. Metrizamide buoyant density gradient centrifugation of chromatin in various ionic environments. 100 μl of chromatin (0.65 mg DNA/ml) was centrifuged in metrizamide gradient solutions in 1 mM Tris-HCl (pH 8.0) alone (A), with 5 mM $MgCl_2$ (B), with 0.14 M NaCl (C) or with 1.0 M NaCl (D). After a centrifugation for 12 h at 35 000 rev/min at 5°C, an aliquot (0.2 ml) of fractions was assayed for density (— — —), protein (○——○), and DNA (●——●). Peak concentrations of chromatin are indicated by arrows, with its buoyant density. Data reproduced from ref. 18 with permission of the authors and publishers.

of salt the buoyant density decreases as a result of the proteins becoming dissociated from the chromatin (*Figure 9*). The presence of millimolar amounts of $MgCl_2$ also increases the buoyant density of chromatin in metrizamide gradients (*Figure 9*). In both cases the increases in buoyant density of chromatin in the presence of NaCl and $MgCl_2$ would appear to be a reflection of the lower solubility and higher degree of aggregation in these ionic environments. Hence chromatin is less hydrated in these ionic environments. As might be expected, there is a smaller effect of loading on the buoyant density of chromatin when the chromatin is aggregated (18). The observed changes in the degree of aggregation of chromatin as a result of concentration or the presence of ions has been found to be fully reversible (18).

5.2. Purification of Chromatin

Usually chromatin is prepared from purified nuclei (7). However, in cases where it is difficult to obtain purified nuclei, an alternative procedure is to use metrizamide or Nycodenz gradients to prepare pure chromatin from nuclei which are not completely pure. An example of such a method is the technique for preparing yeast chromatin from yeast nuclei (19) since even nuclei obtained from protoplasts are usually contaminated with membrane and cell wall material. *Table 5* gives the protocol for the purification of yeast chromatin using metrizamide gradients. In this case it is important to include the protease inhibitor phenylmethylsulphonyl fluoride (PMSF) in all buffers during the preparation of nuclei and chromatin, otherwise the results obtained are not consistent as a result of the variable loss of proteins from the chromatin resulting from protease activity. The chromatin from active and starved cells bands over the range 1.18 – 1.20 g/ml as a single peak.

Table 5. Purification of Chromatin from Yeast Nuclei.

1. Harvest the cells by centrifugation at 1000xg for 20 min. Wash the cells by resuspension in distilled water and recentrifuge.
2. Suspend the cells in 3 vols of 25 mM EDTA, 0.1 M 2-mercaptoethanol, 20 mM Tris-HCl (pH 8.0) and incubate the cells at 30°C for 20 min. Pellet the cells by centrifugation at 2500xg for 5 min.
3. Resuspend the cells in 3 vols of 0.9 M sorbitol, 1 mM EDTA, 20 mM imidazole-HCl (pH 6.4) and add purified snail gut enzyme (1.0 ml/5 g of cells) and incubate at 30°C until sphaeroplasting is complete (30 – 60 min). Add an equal volume of ice-cold 1.0 M sorbitol and pellet the sphaeroplasts by centrifugation at 2500xg for 10 min at 5°C.
4. Suspend the cells in 0.6 M sorbitol in buffer A (0.5 mM $MgCl_2$, 0.5 mM PMSF, 8% polyvinylpyrrolidone, 0.4% Triton X-100, 20 mM potassium phosphate (pH 6.5) and homogenise the cells by passing the cell suspension once through a 26 gauge needle; this should break about 80% of the cells.
5. Add an equal volume of buffer A to the homogenate and centrifuge the homogenate at 500xg for 5 min to remove whole cells and cell debris; the nuclei remain in the supernatant.
6. Prepare a 15 – 35% (w/v) metrizamide (Nyegaard & Co., Oslo, Norway) gradient containing 0.1 M NaCl, 1 mM EDTA, 1 mM dithiothreitol, 0.5 mM PMSF and 10 mM Tris-HCl (pH 7.5) underlayered with a cushion of 55% (w/v) metrizamide in the same buffer.
7. Prior to loading the sample, overlayer the gradient with 0.7% sucrose, 0.1 M NaCl, 2 mM $CaCl_2$, 1 mM dithiothreitol, 0.7% Brij 58, 1% bovine serum albumin, 0.5 mM PMSF, 10 mM Mops-NaOH (pH 7.4) containing 13 units of bee venom phospholipase A_2.
8. Load the yeast nuclei from Step 5 suspended in 0.3 M sorbitol, 8% polyvinylpyrrolidone, 0.5 mM $MgCl_2$, 0.4% Triton X-100, 0.5 mM PMSF, 20 mM potassium phosphate (pH 6.5) onto the gradient and centrifuge the gradients at 500xg for the first 15 min, then increase the centrifugal force to 180 000xg and continue centrifugation for a further 18 h at 5°C.
9. The chromatin bands at 1.18 – 1.20 g/ml and can be unloaded by upward displacement.

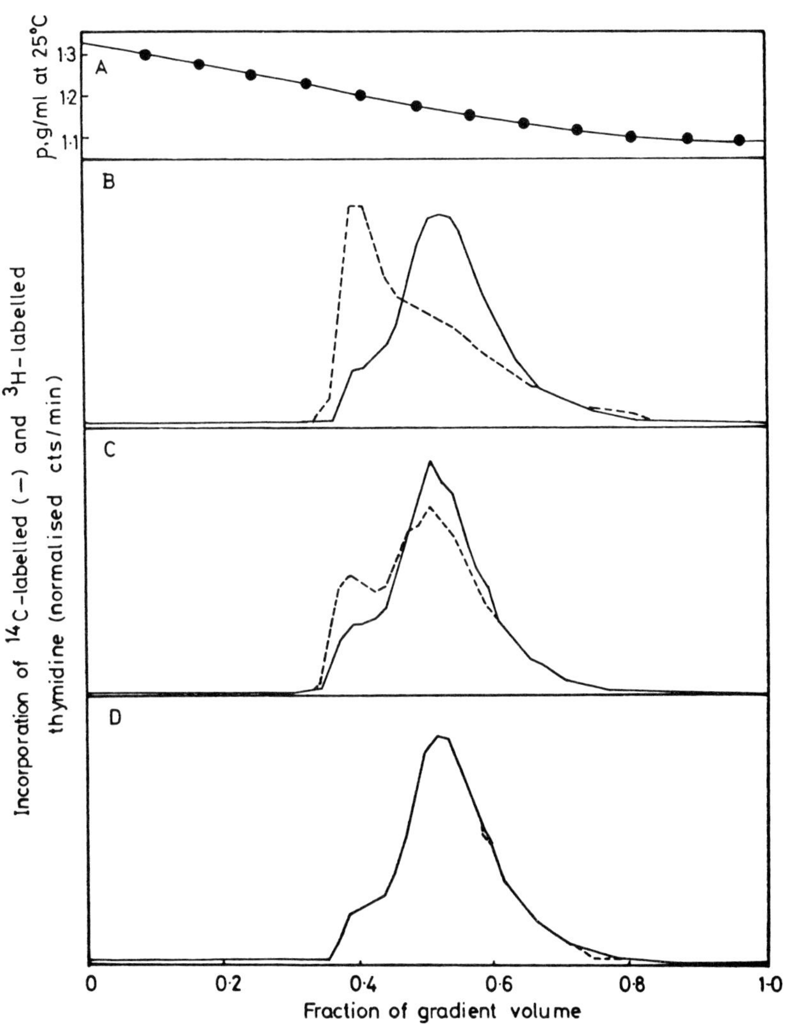

Figure 10. Separation of newly replicated chromatin from bulk chromatin in metrizamide gradients (A). Cells were labelled for 24 h with ^{14}C-thymidine (———) and then for 1 min (B), 10 min (C), or 100 min (D) with ^{3}H-thymidine (— — —). After digestion with micrococcal nuclease at 5.33 x 10^7 units/nucleus and removal of acid-soluble material (2.4% ^{14}C; 8.8, 2.4, 2.4% ^{3}H), the total nucleosomal fraction was centrifuged in metrizamide gradients. Data derived from ref. 25.

Chromatin from yeast cells treated with lomofungin, an inhibitor of RNA synthesis; bands as a dense aggregate at 1.27 g/ml.

This protocol, with minor modifications, should be suitable for a wide range of nuclei which are difficult to purify. Careful choice of the amount of chromatin loaded onto metrizamide gradients will allow the chromatin to be banded at a density such that it does not coband with other possible contaminants. This type of separation has also been used to purify the nascent RNA associated with chromatin from the rest of the nuclear RNA (20).

Metrizamide gradients have also been used as a method for preparing small amounts of chromatin, especially from the tissue culture cells, without the usual large losses of material that occur when most of the standard methods for preparing chromatin are used (21). Similarly, metrizamide gradients have been used for the preparation of viral nucleoprotein structures of SV40 (22); interestingly enough in this study no differences were found between the replicative and mature forms of the virus.

5.3. **Fractionation of Chromatin**

5.3.1. *Replicating Chromatin*

During replication, changes occur in the structure and composition of chromatin close to the replication fork (for a review see ref. 23). When chromatin from dividing cells is fractionated on metrizamide gradients a small amount of the chromatin bands at a marginally higher density than the bulk of the chromatin. Pulse labelling with thymidine reveals that the denser chromatin is associated with the newly replicated DNA (*Figure 10*). The similarity of the fractionations of replicating chromatin obtained with such varied cell types as sea urchin embryos (24) and a line of tissue culture cells derived from mouse leukaemia cells (25) suggests that isopycnic fractionation in metrizamide or Nycodenz gradients may prove to be a general method for the isolation of fractions enriched in replicating chromatin.

Other investigations have used alkaline metrizamide gradients to investigate the deposition of histones on newly replicated chromatin (26). The method used is to denature the chromatin at pH 12.5, the strands of DNA separate but the histones remain associated with the DNA. The denatured chromatin is then loaded onto a self-forming gradient of 27% (w/v) metrizamide dissolved in 20% (v/v) triethanolamine giving a final pH of pH 10.5. The use of triethanolamine instead of NaOH to adjust the pH of the gradient avoids the possibility of the histones dissociating and becoming redistributed along the DNA.

5.3.2. *Sonicated Chromatin*

Sonication of chromatin causes random fragmentation of the DNA. High-molecular weight chromatin forms a single band in metrizamide gradients with a density of up to 1.20 g/ml depending on the loading (see Section 5.1). After mild sonication, 5 sec or less, the size of the DNA is reduced to 800 bp, though the chromatin still forms a single band (27). Further sonication reduces the size of DNA to a minimum of 300 bp after sonication for 2 min (*Figure 11*). Prolonged sonication of the chromatin leads to the formation of protein-rich chromatin fragments which can be separated from the bulk of the chromatin on metrizamide gradients (*Figure 11*); the buoyant densities of the main and satellite bands are 1.18 g/ml and 1.24 g/ml, respectively. Similar separations can also be obtained using Nycodenz gradients. Such fractionations are very reproducible and similar fractionations can be achieved using chromatin from a number of different tissues. However, there appears to be no detectable fractionation of DNA

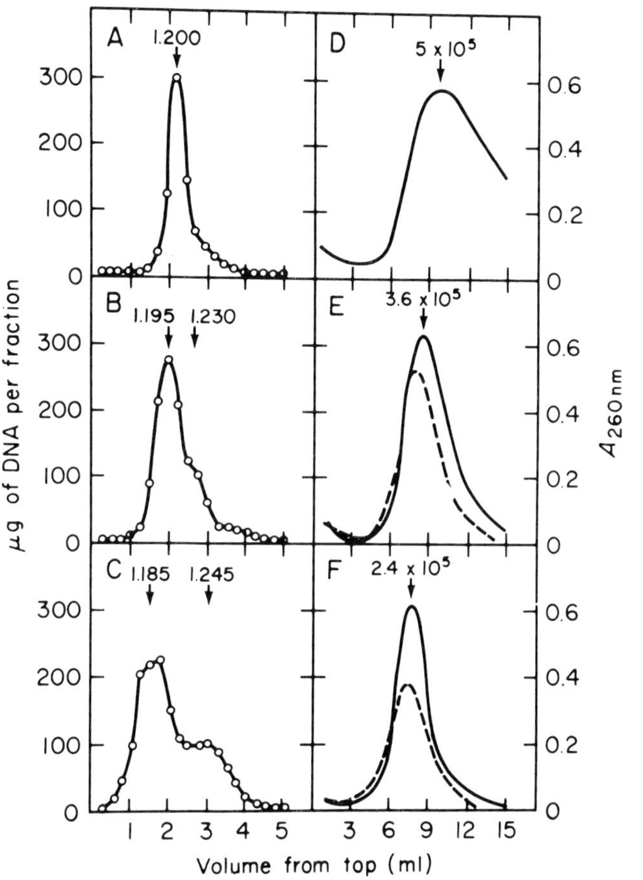

Figure 11. The effect of shearing on the banding pattern of adult mouse-liver chromatin in metrizamide gradients. Chromatin was prepared from mouse liver, sheared by sonication for 5 sec (A), 30 sec (B), or 2 min (C), then centrifuged to equilibrium in metrizamide gradients (○——○) in an MSE 10 x 10 ml rotor at 60 000xg for 44 h at 2°C. Buoyant densities (g/ml) (arrows) were calculated from refractive indices. DNA was isolated from the light and dense fractions separated from chromatin which had been sonicated for 5 sec (D), 30 sec (E), or 2 min (F) and sedimented through neutral 5–11% (w/w) sucrose gradients. The arrows indicate the peak molecular weights of the DNA; (———) DNA from the light chromatin fraction; (— — —) DNA from the dense chromatin fraction. Data reproduced from ref. 27 with permission of the authors and publishers.

sequences in this fractionation and the significance of this type of fractionation, if any, remains unknown.

5.3.3. *Nuclease-digested Chromatin*

Nuclease digestion of chromatin preferentially digests particular regions of the chromatin. Digestion of nuclei with DNase I selectively digests the active chromatin while digestion with staphylococcal nuclease or DNase II preferentially digests the linker DNA between nucleosomes.

Surprisingly, little work has been done on the banding of nuclease-digested

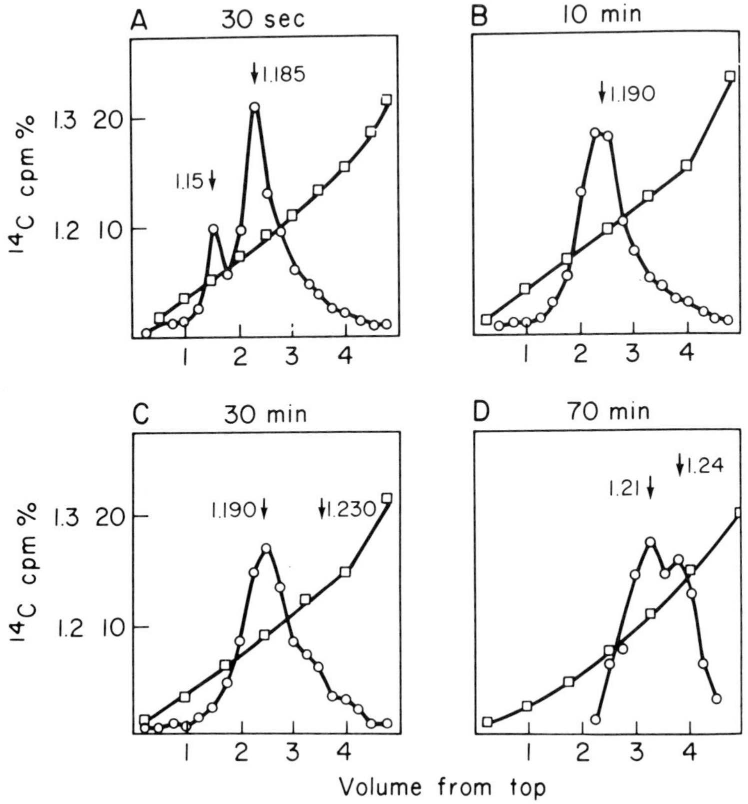

Figure 12. Isopycnic banding of M2-cell chromatin after digestion with staphylococcal nuclease. (A) 30 sec; (B) 10 min; (C) 30 min and (D) 70 min. Nuclei were prepared from cells grown with ^{14}C-thymidine in the culture medium. The digestion mixture contained 1 mM $CaCl_2$, 5 mM Tris-HCl (pH 8.5) staphylococcal nuclease 0.5 μg/ml and a final chromatin-DNA concentration of 300 μg/ml. Tubes containing 5 ml of 36% (w/v) metrizamide solution in 1 mM EDTA, 1 mM Hepes-NaOH (pH 7.0) were centrifuged at 100 000xg for 42 h at 2°C. The density profile (□—□) and the distribution of DNA (○——○) in the gradients were measured. Data derived from ref. 28.

chromatin in metrizamide gradients (19, 25, 28). Mild digestion of chromatin leads to the formation of a low density component (*Figure 12*); however, this material rapidly disappears and the chromatin becomes denser as the DNA is digested (*Figure 12*). Again there would seem to be no fractionation of DNA sequences at least using the method described by these workers.

6. ISOLATION OF NUCLEOIDS

As described in the Introduction, in bacteria and the mitochondria and chloroplasts or eukaryotic cells the DNA is present in the form of a single chromosome attached to the inner membrane. Whilst traditionally bacterial nucleoids have been isolated by rate-zonal centrifugation on sucrose gradients, interest in the methods for the isolation of nucleoids from eukaryotic organelles has been more recent and so attempts have been made to use metrizamide gradients for the isolation of nucleoids of organelles. A wide variety of mitochondrial

Table 6. Preparation of Purified Nucleoids from Yeast Mitochondria.

1. Prepare purified mitochondria using one of the standard methods (34) as described in ref. 4.
2. Suspend the mitochondria in 0.14 M NaCl, 1 mM EDTA, 0.5 mM PMSF, 10 mM Tris-HCl (pH 7.4) by gentle hand homogenisation to give a final concentration of 2 mg/ml of protein.
3. Incubate the mitochondrial suspension at 25°C and add a one tenth volume of 20% (v/v) Triton X-100. Mix thoroughly, add 25 μg of boiled pancreatic RNase for each 100 mg of protein and continue the incubation for another 60 sec.
4. Cool the mitochondrial lysate to 0°C and load the lysate (approx. 4 mg of protein) onto a 10 ml preformed gradient of 20% – 55% (w/v) metrizamide (Nyegaard & Co., Oslo, Norway) containing 0.14 M NaCl, 1 mM EDTA, 10 mM Tris-HCl (pH 7.4).
5. Centrifuge the gradients at 100 000xg for 18 h at 5°C; smaller nucleoids require longer centrifugation times.
6. Yeast nucleoids band as a white aggregate at 1.26 g/ml.

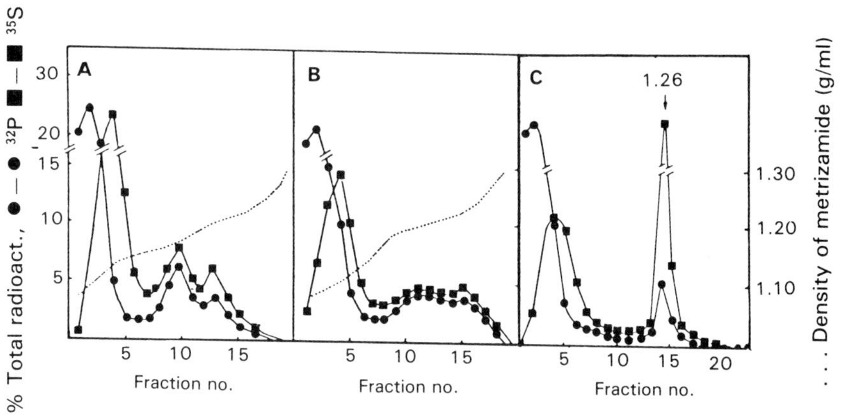

Figure 13. Effect of lysis conditions on the banding patterns of yeast mitochondrial nucleoids on metrizamide gradients. Yeast mitochondria, suspended in 0.14 M NaCl, 10 mM Tris-HCl (pH 7.4), 1 mM EDTA, were lysed with 2% Triton X-100 at either 0°C (A) or at 30°C (B,C) in the absence (B) or presence (C) of ribonuclease. The lysed mitochondria were centrifuged in preformed metrizamide gradients of the same ionic composition as described in *Table 6*. The distributions of ^{35}S-labelled proteins (——) and ^{32}P-labelled nucleic acids (● —— ●) were measured in each case.

nucleoids have been isolated using metrizamide gradients (4, 29, 30) and the yields of nucleoid are much higher than those obtained using rate-zonal centrifugation methods.

The conditions for the purification of yeast mitochondrial nucleoids have been thoroughly investigated and *Table 6* gives the experimental procedure that should be used. A notable feature of this procedure is that it is necessary to digest the mitochondrial lysate with RNase to minimise membrane contamination, presumably because the nucleoid is attached to membrane material by nascent RNA. In addition, in the case of yeast mitochondria, the lysis of the mitochondria must be carried out at 30°C, otherwise the lysis of the mitochondria by Triton X-100 is inefficient and large amounts of membrane remain associated with the nucleoid structure decreasing the buoyant density of the complex (*Figure 13*). It may be that yeast mitochondria are unusual in requiring such a high lysis temperature, since experiments with mitochondria of rat liver and *X. laevis* liver suggest that in these cases efficient lysis is possible at 5°C.

The buoyant density of mitochondrial nucleoids in metrizamide gradients is dependent on the amount of nucleoid loaded onto the gradient. As in the case of chromatin (see Section 5.1), increasing the amount of nucleoid loaded onto the gradient increases the buoyant density up to a maximum of 1.26 g/ml. Hence it is possible to ensure that the nucleoids band at a density completely separate from possible contaminants, particularly membrane material. Since the buoyant density of nucleoids can be manipulated to a lesser extent in Nycodenz gradients (Section 5.1), it would appear that, in this case, it may be more appropriate to use metrizamide gradients rather than Nycodenz gradients.

7. REFERENCES

1. Igo-Kemenes,T., Horz,W. and Zachau,H.G. (1982) *Ann. Rev. Biochem.*, **51**, 89.
2. Small,D., Nelkin,B. and Vogelstein,B. (1982) *Proc. Natl. Acad. Sci. (USA)*, **79**, 5911.
3. Rouviere-Yaniv,J., Yaniv,M. and Germond,J.E. (1979) *Cell*, **17**, 265.
4. Chambers,J.A.A., Rickwood,D. and Barat,M. (1981) *Expl. Cell Res.*, **133**, 1.
5. Smuckler,E.A., Koplitz,M. and Smuckler,D.E. (1976) in *Subnuclear Components: Preparation and Fractionation* (Birnie,G.D. ed.) Butterworths, London, p. 1.
6. Wray,W. and Stubblefield,E. (1970) *Expl. Cell Res.*, **59**, 469.
7. Rickwood,D. and Birnie,G.D. (1976) in *Subnuclear Components: Preparation and Fractionation* (Birnie,G.D. ed.) Butterworths, London, p. 129.
8. Wray,W. (1976) in *Biological Separations in Iodinated Density Gradient Media* (Rickwood,D. ed.) IRL Press Ltd., Oxford and Washington, p. 57.
9. Wortman,M.J. and Segal,A. (1981) *Anal. Biochem.*, **115**, 147.
10. Risley,M.S., Gambino,J. and Eckhardt,R.A. (1979) *Develop. Biol.*, **68**, 299.
11. Mathias,A.P. and Wynter,C.V.A. (1973) *FEBS Lett.*, **33**, 18.
12. Meistrich,M.L. (1977) in *Methods in Cell Biology* (Prescott,D.M. ed.) Academic Press Inc., New York, Vol. **15**, p. 16.
13. Chambers,J.A.A. and Rickwood,D. (1978) in *Centrifugation: A Practical Approach* (Rickwood, D. ed.) IRL Press Ltd., Oxford and Washington, p. 33.
14. Higashinakagawa,T., Sezaki,M. and Kondo,S. (1979) *Develop. Biol.*, **69**, 601.
15. Higashinakawaga,T., Wahn,H. and Reeder,R.H. (1977) *Develop. Biol.*, **55**, 375.
16. Stubblefield,E. (1968) in *Methods in Cell Physiology* (Prescott,D.M. ed.) Academic Press Inc., New York, Vol. **3**, p. 25.
17. Wray,W. (1977) in *Methods in Cell Biology* (Prescott,D.M. ed.) Academic Press Inc., New York, Vol. **15**, p. 111.
18. Kondo,T., Nakajima,Y. and Kawakami,M. (1979) *Biochim. Biophys. Acta*, **561**, 526.
19. Gullov,K.B. and Friis,J. (1978) *Expl. Mycol.*, **2**, 161.
20. Kimmel,C.B., Sessions,S.K. and MacLeod,M.C. (1976) *J. Mol. Biol.*, **102**, 177.
21. Monahan,J.J. and Hall,R.H. (1975) *Anal. Biochem.*, **65**, 187.
22. Cremisi,C., Pignatti,P.F., Croissant,O. and Yaniv,M. (1976) *J. Virol.*, **17**, 204.
23. Sheinin,R., Humbert,J. and Pearlman,R.E. (1978) *Ann. Rev. Biochem.*, **47**, 277.
24. Levy,A., Jakob,K.M. and Moav,B. (1975) *Nucleic Acids Res.*, **2**, 2299.
25. Murphy,R.F., Wallace,R.B. and Bonner,J. (1980) *Proc. Natl. Acad. Sci. (USA)*, **77**, 3336.
26. Russev,G. and Tsanev,R. (1976) *Nucleic Acids Res.*, **3**, 697.
27. Rickwood,D., Hell,A., Malcolm,S., Birnie,G.D., Macgillivray,A.J. and Paul,J. (1974) *Biochim. Biophys. Acta*, **353**, 353.
28. Malcolm,S. and Paul,J. (1976) in *Biological Separations in Iodinated Density Gradient Media* (Rickwood,D. ed.) IRL Press Ltd., Oxford and Washington, p. 41.
29. Rickwood,D. and Jurd,R.D. (1978) *Biochem. Soc. Transact.*, **6**, 266.
30. Sevaljevic,L., Petrovic,L.S. and Rickwood,D. (1978) *Mol. Cell Biochem.*, **21**, 139.
31. Wray,W., Johnson,J., Gollin,S.M., Barr,C. and Wray,V.P. (1982) *J. Cell Biol.*, **95**, 465a.

32. Yaneva,M. and Dessev,G. (1976) *Eur. J. Biochem.,* **66**, 535.
33. MacGillivray,A.J. and Rickwood,D. (1978) in *The Cell Nucleus* (Busch,H. ed.) Academic Press Inc., New York, San Francisco, London, Vol. **6**, p. 263.
34. Chappell,J.B. and Hansford,R.G. in *Subcellular Components: Preparation and Fractionation* (Birnie,G.D. ed.) Butterworths, London, p. 77.

CHAPTER 5

The Fractionation and Subfractionation of Cell Membranes

J. GRAHAM, D. BAILEY, J. WALL, K. PATEL and S. WAGNER

1. INTRODUCTION

1.1. Domains in the Plasma Membrane

The existence of compositionally-distinct domains at the surface of cells from organised tissues such as liver, intestinal mucosa and kidney proximal tubule is well established. Hepatocytes, for example, exhibit three morphologically and functionally distinct surfaces, namely the blood sinusoidal surface, the bile canalicular surface and the contiguous membrane. These three plasma membrane domains have been isolated and characterised (1). The bile canalicular membrane is enriched in alkaline phosphatase, 5′-nucleotidase and leucine aminopeptidase, the blood sinusoidal membrane in glucagon-stimulated adenyl cyclase and the contiguous membrane in Na^+/K^+-ATPase. Enterocytes (the mucosal cells lining the lumen of the intestine) show a clear surface morphological polarity: adjacent to the intestinal lumen is the brush-border membrane which consists of many microvilli and the remainder of the surface which is called the basolateral membrane. The former is heavily enriched in alkaline phosphatase, maltase, leucine aminopeptidase and other peptidases; the latter is enriched in Na^+/K^+-ATPase (2–4). The microvillar membrane of the kidney proximal tubule cells shows a similar enrichment of peptidases and alkaline phosphatase similar to that of the intestine, while, as in the case of liver, the Na^+/K^+-ATPase is enriched in the basolateral membrane (5). In kidney and liver, the 5′-nucleotidase tends to co-purify with the alkaline phosphatase and peptidases (1,5) while in the enterocyte the 5′-nucleotidase apparently co-purifies with the Na^+/K^+-ATPase (3).

1.2. Heterogeneity of Endomembranes

Synthesis of the components of the plasma membrane occurs almost exclusively within the cytoplasm on elements of the endoplasmic reticulum and Golgi membranes. The transfer of synthesised glycoproteins and proteins from the endomembranes to the surface membrane, particularly that of the integral proteins is generally accepted as occurring by way of a membrane vesicle which pinches off from the Golgi and then fuses with the plasma membrane (6). The existence of compositionally-distinct domains at the surface adds a complication to the membrane synthetic and vesicle transport systems of the cell, in that specific proteins must be transported to specific parts of the surface. The process has been envisaged by Evans (7) to occur by one of three possible pathways. In the case of the hepatocyte he considered three types of proteins: (a) characteristic of the blood sinusoidal membrane, (b) characteristic of the contiguous membrane and (c) characteristic of the bile canalicular membrane. Three possible routes of insertion are proposed: random insertion in which vesicles bearing all three types of proteins fuse randomly with the plasma membrane followed by diffusion of proteins through the plane of the membrane to their final locations; sinusoidal insertion which involves a similar migration of proteins after fusion of a vesicle bearing all three types of proteins at the sinusoidal surface; and domain-specific insertion in which specific vesicles bearing one type of protein fuse specifically with the appropriate membrane domain. A fourth possibility, a variant of the sinusoidal insertion, could involve the redistribution of specific proteins from the sinusoidal membrane via an endocytosed membrane vesicle which moves through the cytoplasm to fuse with, for example, the bile canalicular membrane; this route may indeed form an important part of the blood-bile transport system. In a non-dividing tissue such as adult liver, the flow of membrane to the sinusoidal surface which occurs as part of the active secretory activity of the liver, must be balanced by an equal inward flow. Multiple routes are now recognised all involving endocytosis and the fate of the endocytosed vesicle would appear to depend on its composition. The vesicles may fuse with lysosomes, where the protein and lipid components are broken down and subsequently reutilised by the rough endoplasmic reticulum. They may also fuse with the Golgi membranes and then return to the surface. Alternatively, they may return, after some enzymatic modification, directly to the surface or transfer to the bile canalicular membrane as described above. These routes have been discussed by Evans (7).

1.3. Isolation of Membrane Subfractions by Centrifugation

For the isolation of 'large scale' surface membrane domains such as brush-border membranes, basolateral membranes from kidney and intestine and contiguous membranes from liver, use can be made of the fact that under gentle homogenisation conditions, structures within the cell such as the terminal web of the brush border and junctional complexes between the membranes of adjacent cells tend to maintain these structures as large membrane fragments, making their resolution from other membrane structures relatively easy in sucrose gradients. Not only are they large structures and therefore much more rapidly sedimenting than the membrane vesicles, but also the large amounts of non-membrane proteins associated with these structures tend to make them relatively dense. A large fragment of plasma membrane may be up to 20 μm in diameter and have a density of 1.15 – 1.19 g/ml. The corresponding values for a typical plasma membrane vesicle are 0.05 – 3 μm and 1.07 – 1.19 g/ml, respectively (8). Vesicles from the smooth endoplasmic reticulum have roughly the same dimensions and density as the plasma membrane vesicles (8) but nevertheless sucrose gradients have been successful in resolving rat liver sinusoidal plasma membrane vesicles from smooth endoplasmic vesicles (1).

The subfractionation of endomembrane vesicles is less well understood. Many fractionations of rough and smooth microsomes from rat liver in sucrose gradients have shown that different enzymes are enriched in different regions of the gradient (for a review of this work see ref. 8). There are rather few instances in which a specific endomembrane vesicle has been isolated at a preparative level. One example of this is the diacytosome (9), a vesicle derived from the sinusoidal surface of the hepatocyte which returns to the surface without fusion with any other internal membrane. Diacytosomes have been isolated using a combination of sucrose gradient centrifugation and free flow electrophoresis (10). However, a more thorough investigation into this complex vesicular traffic requires the application of a wide range of modern fractionation techniques, one of which is the use of iodinated density-gradient media.

Metrizamide and Nycodenz, both from Nyegaard & Co. (Oslo, Norway), have been used to separate membrane fractions and subfractions from cell homogenates in situations where sucrose gradients have proved unsatisfactory. The properties of solutions of metrizamide and Nycodenz (11,12), that is the much lower osmotic activity and the lower viscosity compared to solutions of sucrose of the same density, mean that the behaviour of osmotically-active membrane-bound organelles and membrane vesicles will be significantly different in gradients of metrizamide and Nycodenz compared to their behaviour in sucrose gradients. In gradients of metrizamide and Nycodenz, under the appropriate loading conditions, mitochondria, lysosomes and peroxisomes can be separated (13,14); metrizamide gradients have been used to fractionate rough and smooth microsomes (15) and membrane vesicles from Lettree cells (16).

The efficacy of iodinated density-gradient media in subfractionating plasma membranes and endomembranes will be demonstrated with reference to the fractionation of a crude guinea pig enterocyte mitochondrial pellet and the subfractionation of smooth and rough microsomes from rat liver. In both of these systems emphasis will be placed on the ability of metrizamide and Nycodenz to demonstrate the existence of domains both at the surface of the cell, in particular to extend our knowledge of the complexity of this phenomenon at the surface of the enterocyte, and on endomembranes, and also to demonstrate the existence of well-defined areas of the rough and smooth endoplasmic reticulum and the Golgi membranes of rat liver which may be involved in different synthetic processes.

It should be pointed out that the enterocyte system was chosen because of the difficulty in resolving discrete, well-defined fractions using sucrose gradients and because of our interest in isolating a particular plasma membrane subfraction from it. Particle aggregation problems due to the presence of mucus in enterocyte fractions are not encountered with material from other tissues; by choosing the appropriate gradient and centrifugation conditions, however, this problem can be overcome. The optimal experimental protocol for systems used by other workers, even those using the same tissue, may require modification from those detailed in the following sections for a number of reasons, but primarily because homogenisation conditions are virtually impossible to standardise from one laboratory to another. Thus the size of membrane fragments will vary with the method of homogenisation and consequently so will gradient separations based on particle size. The aim of this chapter is to emphasize the advantages of using iodinated density-gradient media over the more routinely used sucrose gradients. After presentation of the results with the guinea pig enterocyte mitochondrial fraction and the rat liver microsomal fraction the use of other sources of material will be briefly discussed.

2. PREPARATION AND FRACTIONATION OF GRADIENTS

2.1. Formation of Gradients

All metrizamide and Nycodenz (Nyegaard & Co. Oslo, Norway) concentrations are given in terms of % w/v. Stock solutions of 50% metrizamide or Nycodenz in 5 mM Tris-HCl (pH 8.0) are kept frozen until required. Gradient solutions are prepared by dilution with 5 mM Tris-HCl (pH 8.0) and as necessary, 1.0 M $MgCl_2$ is added to give final concentrations of 0.5 or 5.0 mM as indicated in the text. Linear gradients are prepared either directly using a standard double-chamber gradient former or by allowing discontinuous gradients to diffuse in an upright position for 16 h at 4°C. When a gradient former is used, the distal end of the delivery tube is positioned at the bottom of the centrifuge tube and the gradient delivered low density first.

The majority of fractionations are carried out using 12 ml gradients in an MSE 6 x 14 ml (Ti) swing-out rotor. For the generation of linear gradients by diffusion, 10–30% metrizamide or Nycodenz gradients are formed from 2.5 ml each of 30%, 25%, 20%, 15% and 2 ml 10%; 10–40% gradients are formed from 3 ml each of 40%, 30%, 20% and 10%; 15–35% gradients are formed from 2.5 ml each of 35%, 30%, 25%, 20% and 2 ml 15%.

2.2. Harvesting of Banded Material

Gradients are collected either by upward displacement with Maxidens, using a cone-shaped Perspex collector on the top of the tube (*Figure 1*), or in cases where the banded material is well-defined and well-separated, a flat-tipped metal cannula attached to a 2 ml syringe is used. The density profile of the gradients or the density of a particular fraction is determined using a refractometer (11,12).

Although neither metrizamide nor Nycodenz significantly inhibit a number of membrane-bound enzymes at concentrations less than 10% (12), to avoid any possible interference from the gradient compounds or Mg^{2+} in enzyme assays, all samples, except those to be processed further for gradient subfractionation are diluted threefold with 5 mM Tris-HCl (pH 8.0), sedimented at 100 000xg for 40 min in an MSE 10 x 10 ml (Ti) angle rotor and resuspended in 0.5–1.0 ml 0.25 M sucrose, 5 mM Tris-HCl (pH 8.0) and frozen at –80°C until required. The samples should be rapidly thawed to 4° prior to use and all enzyme assays are detailed in Appendix 1.

3. PREPARATION AND FRACTIONATION OF ENTEROCYTE MEMBRANES

3.1. Method for the Preparation of Guinea Pig Enterocyte Mitochondrial Fractions

All operations should be carried out at 0–4°C. Slit the proximal jejunum from one young adult guinea pig, killed by cervical dislocation, lengthwise and wash it

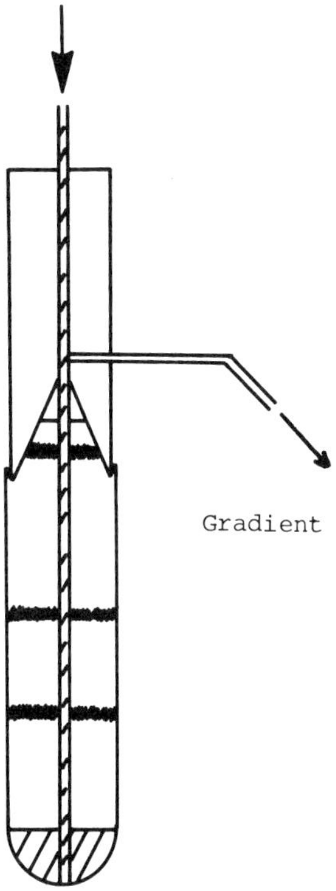

Figure 1. Collection of gradients by upward displacement with Maxidens.

once in isotonic saline and twice in the homogenisation medium (0.25 M sucrose, 1 mM $NaHCO_3$). Using the back of a scalpel blade, gently scrape the mucosa away from the gut wall and transfer it to 30 ml of homogenisation medium. Stir vigorously with a glass rod for about 30 sec to disperse the material. Disruption of the cells is achieved using 4 gentle strokes (up and down) of the pestle of a loose-fitting Dounce homogeniser (Type L, F.T. Scientific Instruments, Station Industrial Estate, Tewkesbury, Glos., UK) followed by 15 strokes of the Teflon pestle (rotating at 500 rev/min) of a Potter-Elvehjem homogeniser (clearance 0.09 mm). The homogenate is then adjusted to 10 mM KCl, 1 mM $MgCl_2$, and 5 mM Tris-HCl (pH 8.0). After checking by phase contrast microscopy that at least 90% of the tissue has been disrupted and that the nuclei are intact, filter the homogenate through a layer of Nylon mesh (pore size 40 μm) to remove any undisrupted tissue and some of the mucus. It is important not to force the homogenate through the mesh but filtration may be aided by gentle agitation.

Under these conditions most of the brush-border remains intact as large fragments although some comminution does occur. These large fragments, nuclei, unbroken cells and cellular debris should be removed by centrifugation at 1000xg for 5 min. The supernatant is carefully decanted and centrifuged at 20 000xg for

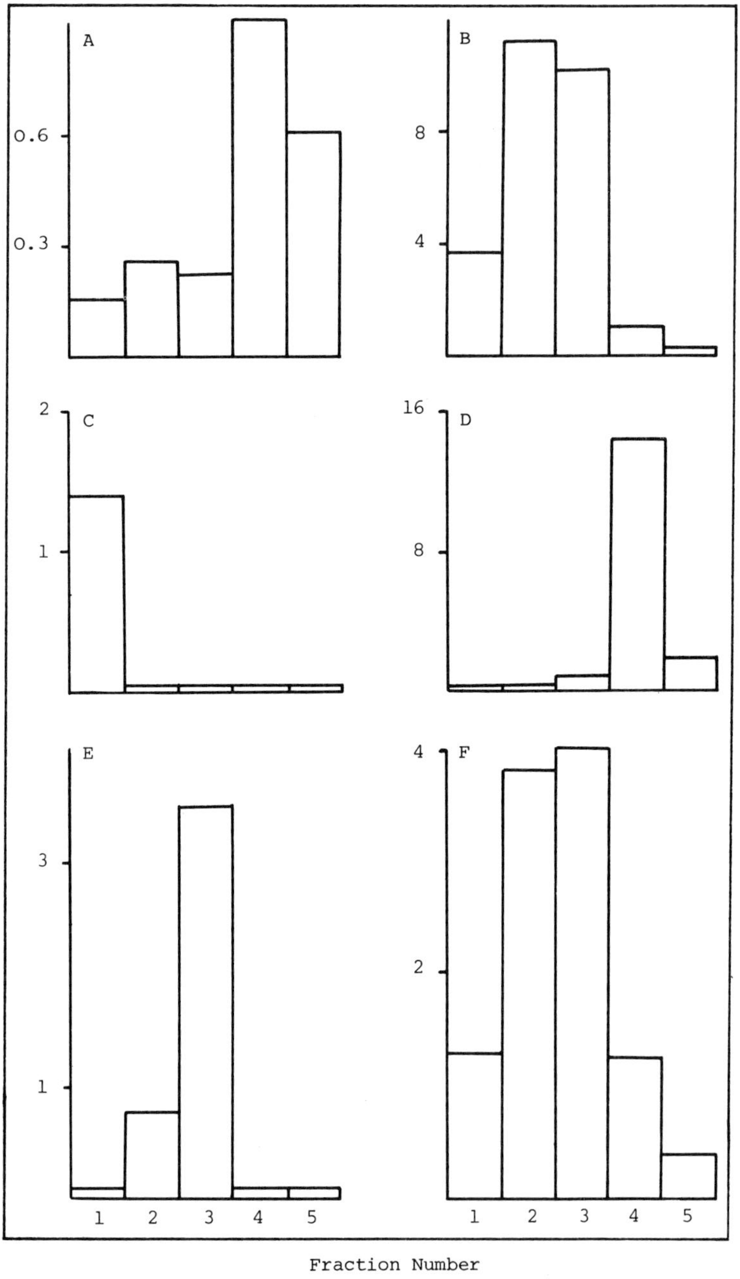

Figure 2. Fractionation of an enterocyte mitochondrial fraction in 10–40% metrizamide gradients: distribution of protein and enzymes. (A) Protein (mg/ml); (B) 5′-nucleotidase; (C) NADPH-cytochrome c reductase; (D) succinate-cytochrome c reductase; (E) ouabain-sensitive Na^+/K^+-ATPase; (F) acid phosphatase. All enzyme activities (ordinates) are given in μmoles substrate utilised/h/mg protein.

10 min in the SS34 rotor of a Sorvall centrifuge. Wash the pellet twice in 30 ml 0.25 M sucrose, 5 mM Tris-HCl (pH 8.0) using the same centrifugation conditions and finally resuspend in 4 ml of this medium by homogenisation using 4-5 strokes of the rotating pestle of the Potter-Elvehjem homogeniser.

3.2. **Fractionation of Enterocyte Mitochondrial Fractions**

3.2.1. *Fractionations in 10–40% Gradients of Metrizamide or Nycodenz*

Load 1 ml of this suspension on to 12 ml 10–40% metrizamide or Nycodenz gradients and centrifuge the gradients at 25 000xg for 60 min in an MSE 6 x 14 ml (Ti) swing-out rotor. Alternatively an HB4 swing-out rotor (Sorvall) with adaptors to take 15 ml tubes can be used at 12 500xg for 2 h. After centrifugation four major bands (bands 2-5) and one minor band (band 1) are observed in both metrizamide and Nycodenz. The enzyme profiles are shown in *Figures 2* and *3*. The majority of the succinate-cytochrome c reductase activity occurs in band 4 in both gradients. The 5′-nucleotidase, aminopeptidase and alkaline phosphatase cosedimented in bands 2 and 3 in both gradients; on the other hand the Na^+/K^+-ATPase which is restricted almost entirely to band 3 in the metrizamide gradient, in Nycodenz is also enriched in band 5. In terms of the total activity the acid phosphatase is widely distributed amongst all the bands, even though bands 2 and 3 contain the highest specific activity. No well-defined lysosome fraction is obtained under these conditions, and furthermore there is evidence that in the enterocyte the acid phosphatase is rather widely distributed amongst various membrane types and not specifically localised in lysosomes (17). It should be noted that the specific activities of the aminopeptidase and alkaline phosphatase in the brush-border fraction, isolated from the nuclear pellet are 25-times and 4-times as high as those in band 3, respectively, while the 5′-nucleotidase activity is very similar. No Na^+/K^+-ATPase activity is detectable in the brush-border fraction. Thus, although the 5′-nucleotidase is not specifically enriched in the brush-border, neither is it associated entirely with the Na^+/K^+-ATPase containing basolateral membranes. This is particularly clear in the Nycodenz fraction enriched exclusively in Na^+/K^+-ATPase.

In sucrose gradients covering the same density range the only clearly defined band contains the mitochondria (*Figure 4*), the remainder of the material is broadly distributed above and below this band. No well-defined regions of the gradient are enriched in acid phosphatase and the 5′-nucleotidase and Na^+/K^+-ATPase activities sediment similarly without any clear cut separation. The absence of well-defined bands makes preparative work in sucrose gradients very difficult.

3.2.2. *Fractionations in 15–35% Metrizamide Gradients*

Load 1 ml of the suspension on to a 12 ml 15–35% metrizamide gradient and centrifuge the gradients at 25 000xg for 60 min in a 6 x 14 ml swing-out rotor. An additional band is resolved above the major mitochondrial band (*Figure 5*). No additional resolution of the major plasma membrane enzymes is obtained, and again a clear cut association of the 5′-nucleotidase with either the alkaline phosphatase and aminopeptidase on the one hand or Na^+/K^+-ATPase on the other is

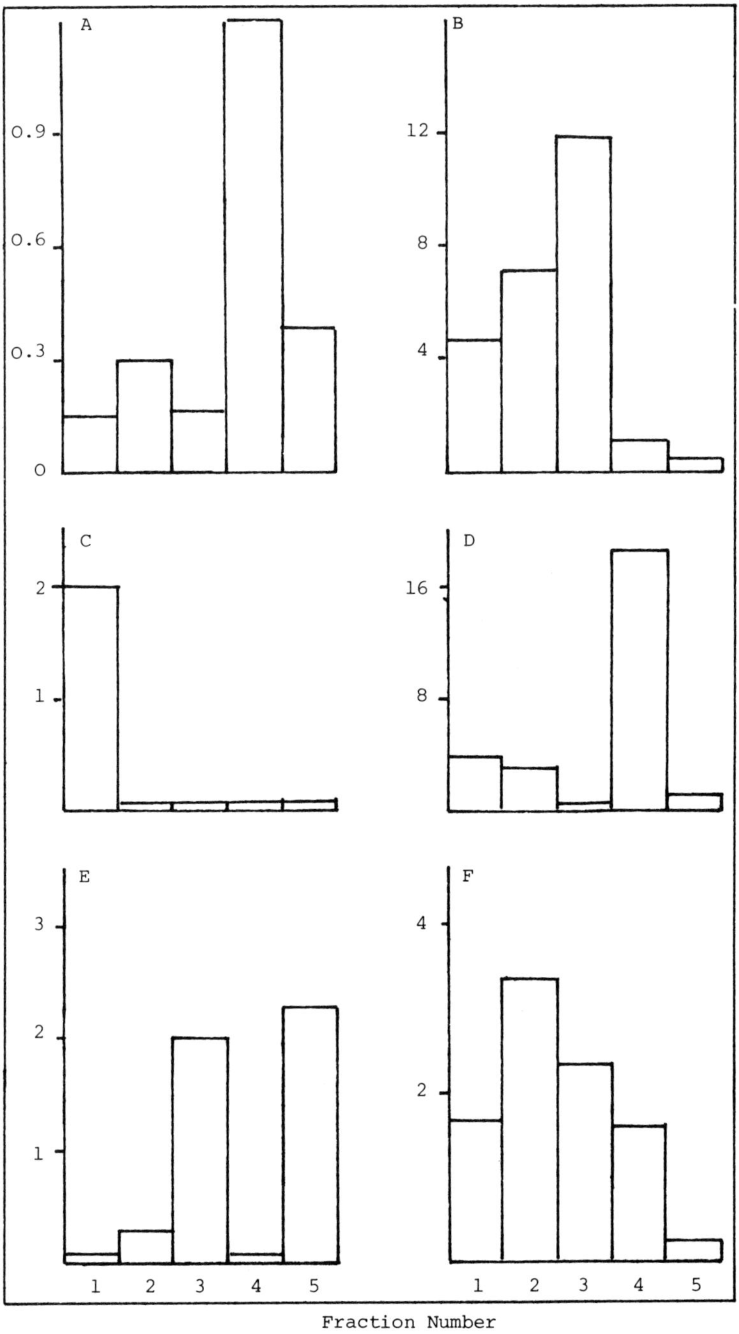

Figure 3. Fractionation of an enterocyte mitochondrial fraction in 10–40% Nycodenz gradients: distribution of protein and enzymes. For details see legend to *Figure 2*.

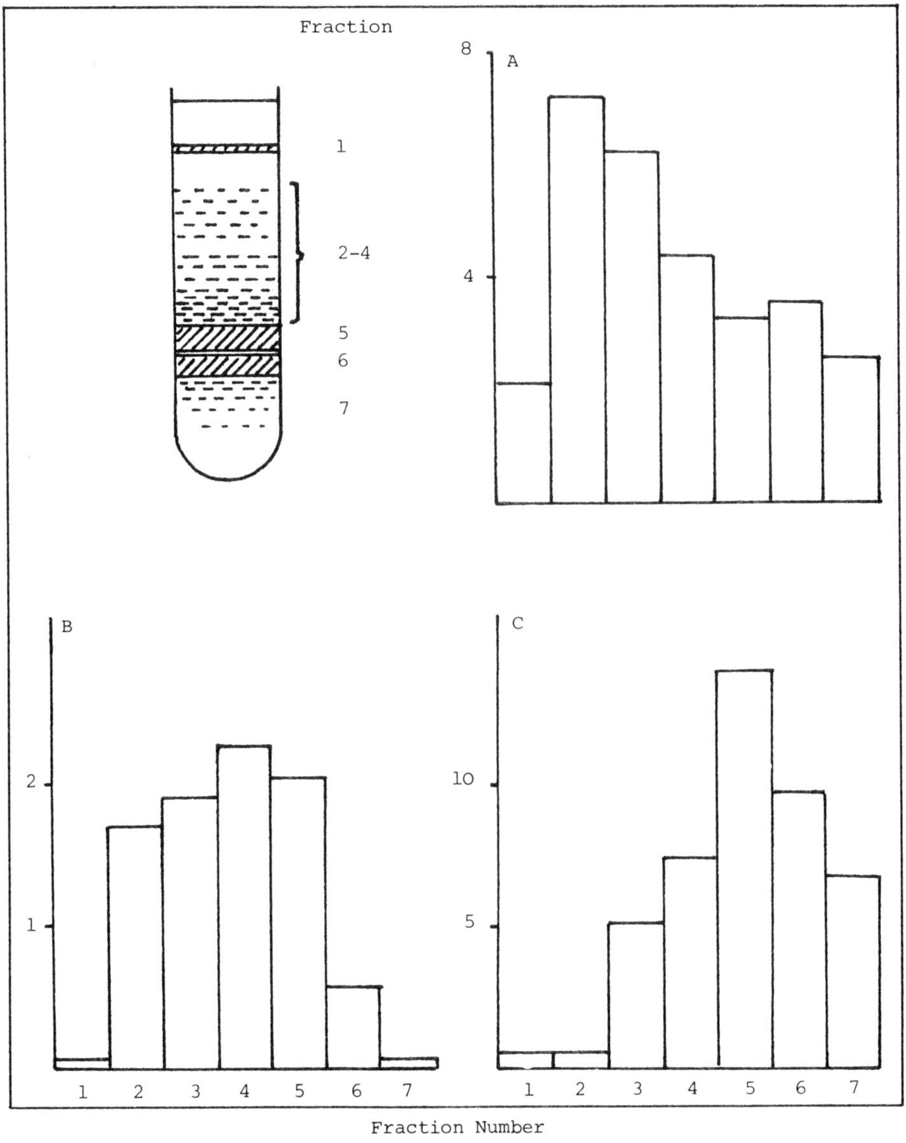

Figure 4. Fractionation of an enterocyte mitochondrial fraction in 15 – 50% (w/w) sucrose gradients: gradient banding pattern and distribution of enzymes. (A) 5′-Nucleotidase; (B) ouabain-sensitive Na^+/K^+-ATPase; (C) succinate-cytochrome c reductase. All enzyme activities (ordinates) are given in μmoles substrate utilised/h/mg protein.

not evident. Although a more well-defined peak of acid phosphatase activity is obtained in band 3, a significant amount of this enzyme still sediments into bands 4 and 5.

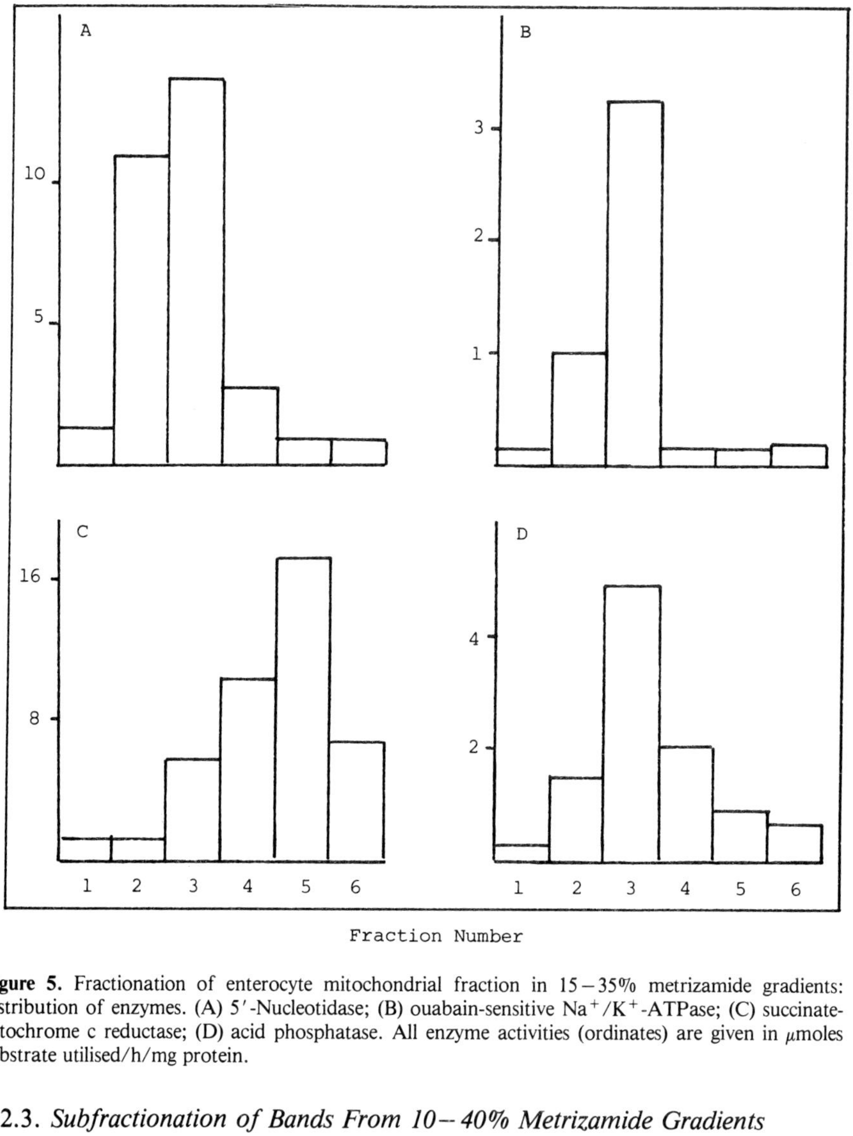

Figure 5. Fractionation of enterocyte mitochondrial fraction in 15–35% metrizamide gradients: distribution of enzymes. (A) 5′-Nucleotidase; (B) ouabain-sensitive Na^+/K^+-ATPase; (C) succinate-cytochrome c reductase; (D) acid phosphatase. All enzyme activities (ordinates) are given in μmoles substrate utilised/h/mg protein.

3.2.3. *Subfractionation of Bands From 10–40% Metrizamide Gradients*

Additional subfractionation of the bands resolved from the mitochondrial fraction in metrizamide can be carried out in shallow secondary gradients.

For further fractionation the major mitochondria-containing band (band 4) should be resuspended in 40% metrizamide, 5 mM Tris-HCl (pH 8.0) and overlayered by 35%, 30%, 25% and 20% (2.5 ml each) and centrifuged at 100 000xg for 16 h in an MSE 6 x 14 ml (Ti) swing-out rotor. Three major bands (bands 3–5) and two minor bands (bands 1 and 2) are obtained (*Figure 6*). This time the acid phosphatase (both in terms of specific activity and total activity) is far better resolved from the major mitochondrial band (band 4). The least dense bands have not been identified but contain small amounts of alkaline phosphatase

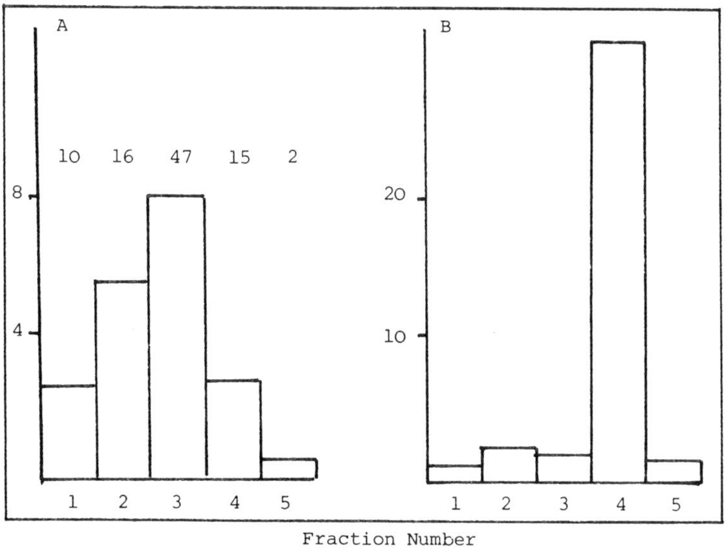

Figure 6. Subfractionation of enterocyte mitochondrial band 4 in bottom-loaded 20 – 40% metrizamide gradients: distribution of enzymes. (A) Acid phosphatase; (B) succinate-cytochrome c reductase. Enzyme activities (ordinates) are given in μmoles substrate utilised/h/mg protein. The figures above the bars in A give the total activity of the acid phosphatase in μmoles phosphate hydrolysed/h.

and 5′-nucleotidase. The efficiency of 'bottom-loading' of a crude mitochondrial fraction in resolving lysosomes and mitochondria has been observed by Wattiaux *et al.* (13,14).

A second type of separation may also be used for the further fractionation of banded material from the 10 – 40% metrizamide gradients. Conditions are chosen so that sedimentation and resuspension of the banded material prior to refractionation is not necessary. The bands are harvested from the first gradient and the concentration of metrizamide in each fraction determined by refractometry. The volume is adjusted with a metrizamide solution of the same concentration, made part of a discontinuous gradient and recentrifuged in an MSE 6 x 14 ml (Ti) swing-out rotor at 100 000xg for 16 h. Discontinuous gradients are set up and centrifuged as described in *Figures 7* and *8*. *Figure 7* shows that the major mitochondrial band (band 4) is resolved into four components; the resolution of the lysosomal and mitochondrial enzyme markers (acid phosphatase and succinate-cytochrome c reductase) is apparently more efficient by this method in which the lysosomes and mitochondria move in opposite directions, than in the bottom-loading method.

Perhaps the most important subfractionation concerns band 3 (*Figure 8*). Four major subfractions are resolved: the splitting of the Na^+/K^+-ATPase activity into a light (band 3-4) and a dense (band 3-2) fraction resembles the dichotomy observed with the crude mitochondrial fraction on Nycodenz gradients. Moreover the lighter fraction is clearly associated with a significant amount of 5′-nucleotidase, whilst the denser fraction is not. These two subfractions (bands 3-2 and 3-4) contained only very low concentrations of alkaline phosphatase. The

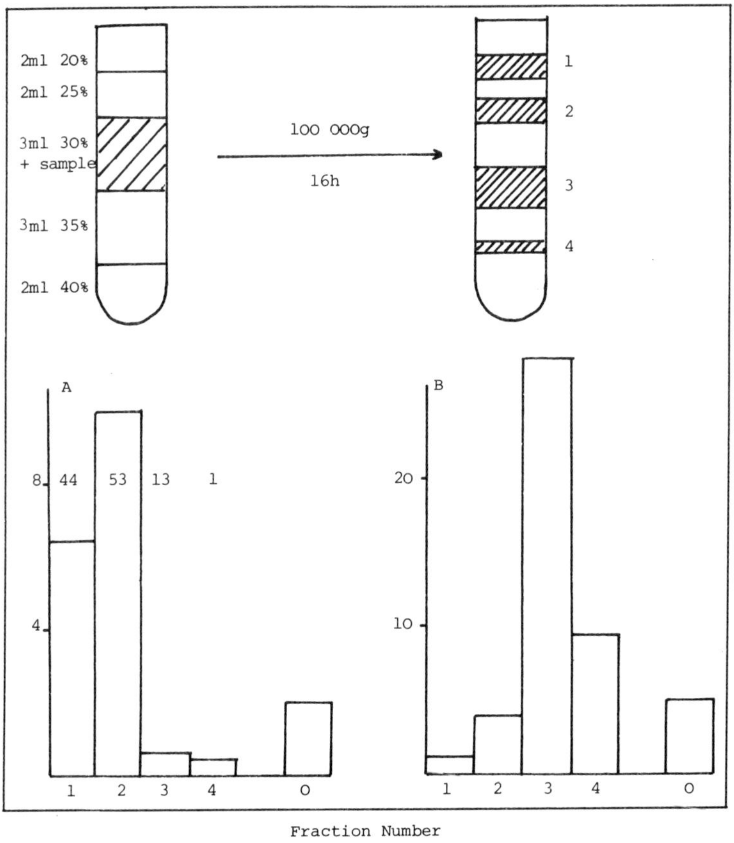

Figure 7. Subfractionation of enterocyte mitochondrial band 4 in median-loaded 20–40% metrizamide gradients: gradient formation; band pattern and distribution of enzymes. Fraction O is the input material. For further details see legend to *Figure 6.*

5′-nucleotidase also exhibited a dichotomy in that while the majority did co-purify with the Na^+/K^+-ATPase (band 3-2), a significant amount of this enzyme co-purifies with the alkaline phosphatase in band 3-1, a fraction which is devoid of Na^+/K^+-ATPase activity.

3.3. Important Practical Points

(i) It is essential that the mitochondrial pellet is washed well in 0.25 M sucrose, 5 mM Tris-HCl (pH 7.4) before application to the gradients. Cations present in the homogenisation medium, important in maintenance of the nuclei in an intact state, must be removed completely. Failure to do this results in aggregation of the mitochondrial band and consequent sequestration of other membranous material.

(ii) The difference in density between the sample and the top of the 10–40% metrizamide or Nycodenz gradient is minimal. This is important for the attainment of the greatest resolution possible in the gradient: the greater the difference in density between the sample and the top of the gradient the more likely it is that

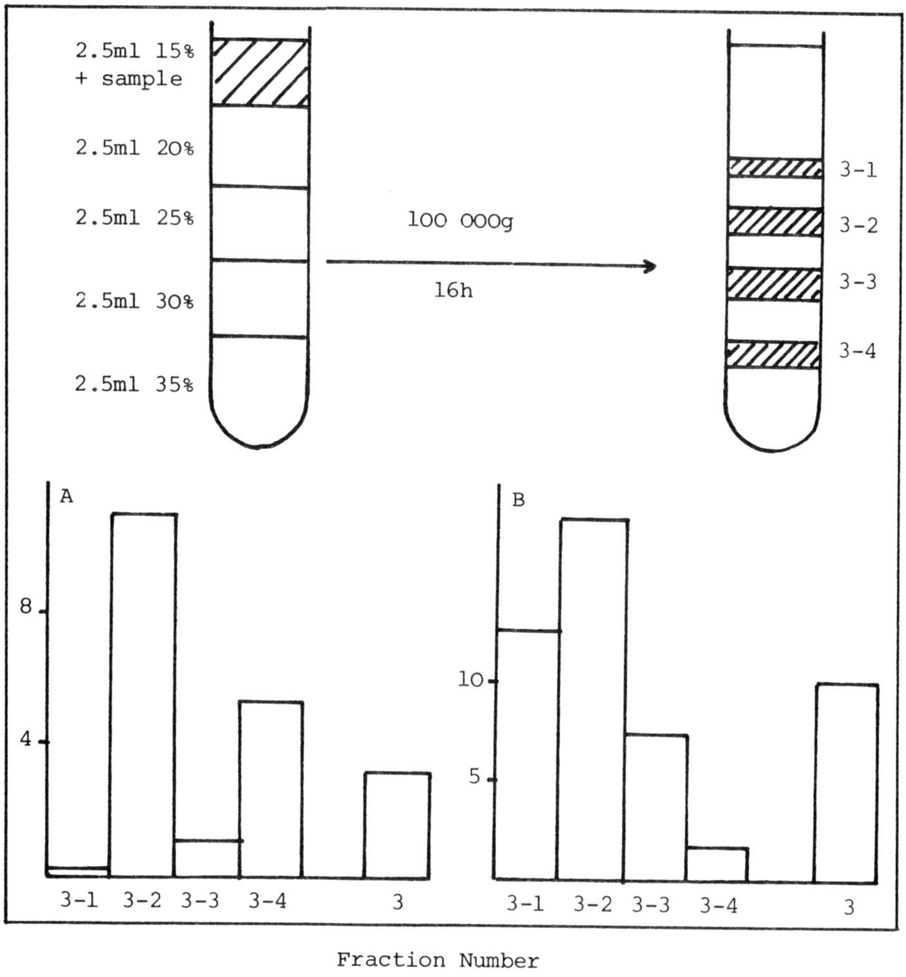

Figure 8. Subfractionation of enterocyte mitochondrial band 3 in 15–35% metrizamide gradients: gradient formation, band pattern and distribution of enzymes. (A) Ouabain-sensitive Na^+/K^+-ATPase; (B) 5′-nucleotidase. Enzyme activities (ordinates) are given in μmoles substrate utilised/h/mg protein.

aggregation will occur. It is thus advisable to check the stability of the interface between the sample and gradient by testing the stability of a small aliquot of the sample on some 10% metrizamide or Nycodenz.

(iii) Excess mucus in the homogenate adversely affects the resolution of these gradients.

3.4 Conclusions

Gradients of metrizamide and Nycodenz are considerably more efficient in resolving the components from an enterocyte mitochondrial fraction than are sucrose gradients. The iodinated density-gradient media are superior both in terms of resolving mitochondria and lysosomes and in resolving subfractions of the plasma membrane. Furthermore, in isopycnic separations of the major mitochondrial band the resolution lysosomes and mitochondria can be improved by loading the sample in the middle of the gradient rather than at the bottom.

The subfractions of the plasma membrane obtainable in iodinated density-gradient media show that the domain organisation in the enterocyte plasma membrane is very complex. While it is clear that the brush-border contains very high concentrations of both aminopeptidase and alkaline phosphatase and that the basolateral membrane is enriched in Na^+/K^+-ATPase, there are further subfractions detectable only by centrifugation in iodinated density-gradient media which indicate the existence of a plasma membrane domain which, like the brush-border, contains alkaline phosphatase but very little aminopeptidase and also contains 5′-nucleotidase, a domain which contains high levels of both 5′-nucleotidase and Na^+/K^+-ATPase and a domain which is apparently enriched solely in Na^+/K^+-ATPase.

4. FRACTIONATIONS OF MEMBRANES FROM OTHER TISSUES

Although the major plasma membrane fragments both from the intestinal mucosal cells and rat liver are sufficiently large to sediment with the nuclei rather than with the mitochondria, a small but significant proportion of the plasma membrane can be recovered from a mitochondrial fraction. In sucrose, the smaller plasma membrane fragments from rat liver are difficult to resolve from the mitochondria because of their similar rates of sedimentation and density. Thus the use of low osmolarity metrizamide and Nycodenz gradients at relatively low centrifugation speeds should benefit the resolution of these components from the mitochondrial fraction from any tissue, although the precise centrifugation conditions, and the gradient composition, such as the level of divalent cations, may require modification. The system may also be applicable to liver hepatoma in which all of the plasma membrane tends to sediment with the mitochondria.

As in the case of rat liver (13,14) top loading of metrizamide gradients with a mitochondria/lysosome fraction gave inferior results when compared with the resolution obtained after bottom loading these particles in metrizamide gradients. Although no data are available, it is likely that the superior resolution obtained by median loading of gradients in the case of the enterocytes may also be achieved with liver.

5. PREPARATION AND FRACTIONATION OF RAT-LIVER MICROSOMES

5.1. Method for the Preparation of Rat-liver Microsomal Fractions

All operations should be carried out at 0–4°C. Kill two adult male rats (approx 250 g) by cervical dislocation and quickly remove their livers into weighed beakers containing approximately 25 ml 0.25 M sucrose, 5 mM Tris-HCl (pH 8.0). Weigh the livers and mince them finely using scissors. Wash the minced liver several times in the sucrose medium to remove as much blood as possible. Resuspend the liver pieces in approximately 1.5-times their volume of medium and homogenise using 4 or 5 strokes of the Teflon pestle (rotating at 500 rev/min) of a

Potter-Elvehjem homogeniser (clearance 0.09 mm). After dilution of the homogenate to 0.25 g of liver wet weight/ml, centrifuge the homogenate at 10 000xg for 20 min in the 8 x 50 ml (SS34) angle rotor of a Sorvall centrifuge. Remove as much of the supernatant as possible using some 2 mm (I.D.) tubing attached to a 50 ml syringe. The supernatant must be aspirated very carefully to avoid disturbing the poorly-packed upper part of the pellet. If necessary recentrifuge the supernatant to eliminate any aspirated pellet material.

To separate the rough and smooth microsomes a modification of the method of Bergstrand & Dallner (18) can be used. Into 10 ml polycarbonate tubes (for the MSE 10 x 10 ml (Ti) fixed-angle rotor) place 3 ml of 1.3 M sucrose, 15 mM CsCl, 5 mM Tris-HCl (pH .80), overlayer it with 1.5 ml of 0.6 M sucrose, 5 mM CsCl, 5 mM Tris-HCl (pH 8.0) followed by 4.5 ml of the 10 000xg supernatant. After centrifugation at 102 000xg for 90 min the smooth microsomes band as a doublet just below the interface between the 0.6 M and 1.3 M sucrose and the rough microsomes form a loose pellet. Discard the upper phase (above the interface) and remove the two microsomal fractions using a flat-tipped metal cannula attached to a syringe. Dilute the two microsome fractions with at least 3 volumes of 0.25 M sucrose, 5 mM Tris-HCl (pH 8.0) and centrifuge at 100 000xg for 30 min. Wash the pellets once, and finally resuspend them in 2–3 ml of this medium. Prior to further fractionation homogenise the pellets in a 5 ml Potter-Elvehjem homogeniser using 10 strokes of the Teflon pestle (clearance 0.07 mm) rotating at 500 rev/min. Complete disaggregation of the membrane pellets is essential for producing high resolution in the subsequent gradients.

5.2. Method for the Subfractionation of the Rough and Smooth Microsomes

Place the rough microsome fraction (1 ml) on top of a linear 10–40% metrizamide or Nycodenz gradient containing 5 mM Tris-HCl (pH 8.0) and 0.5 mM $MgCl_2$, and centrifuge the gradients at 100 000xg for 4 h. Place the smooth microsome fraction (1 ml) on top of a linear 10–30% metrizamide or Nycodenz gradient containing 5 mM Tris-HCl (pH 8.0) and 5 mM $MgCl_2$ and centrifuge the gradient at 160 000xg for 60 min.

5.2.1. *Fractionation of Rough Microsomes*

Five major fractions are resolved from the rough microsomes in 10–40% gradients of both metrizamide and Nycodenz. The overall heterogeneity revealed in the five bands is very similar in both gradients but the degree of resolution obtained using Nycodenz gradients is superior and only the Nycodenz gradient results are shown in *Figure 9*. It is difficult to compare these results with those obtained by other workers who have used sucrose gradients because of the variety of conditions employed. However, it is worth pointing out that Eriksson (19) observed that the banding densities of rough microsomal subfractions containing ATPase, glucose-6-phosphatase and NADPH-cytochrome c reductase increased in that order and that generally speaking most workers (20) have found that the NADH-cytochrome c reductase-rich vesicles or submicrosomal particles from rough endoplasmic reticulum band at a higher density in sucrose than do the NADPH-

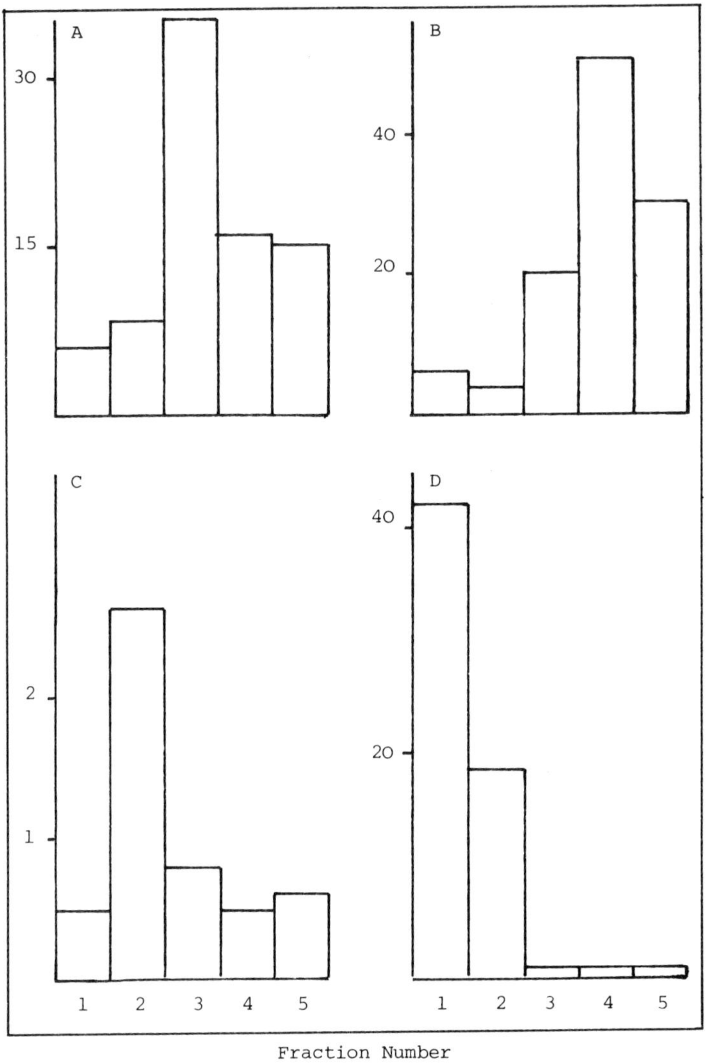

Figure 9. Subfractionation of rough microsomes in 10–40% Nycodenz gradients: distribution of enzymes. (A) NADPH-cytochrome c reductase; (B) NADH-cytochrome c reductase; (C) glucose-6-phosphatase; (D) 5′-nucleotidase. All enzyme activities (ordinates) are given in μmoles substrate utilised/h/mg protein.

cytochrome c reductase-rich fractions. The subfractionations obtained in Nycodenz gradients are in broad agreement with these results. The major advantage of the separations in iodinated density-gradient media is that well-defined, well-separated subfractions are obtained. In contrast, in sucrose gradients the vesicles tend to distribute themselves rather broadly through the gradient and discrete bands are seldom observed. Thus, particularly from a preparative point of view, the separations in Nycodenz and metrizamide represent an important advance.

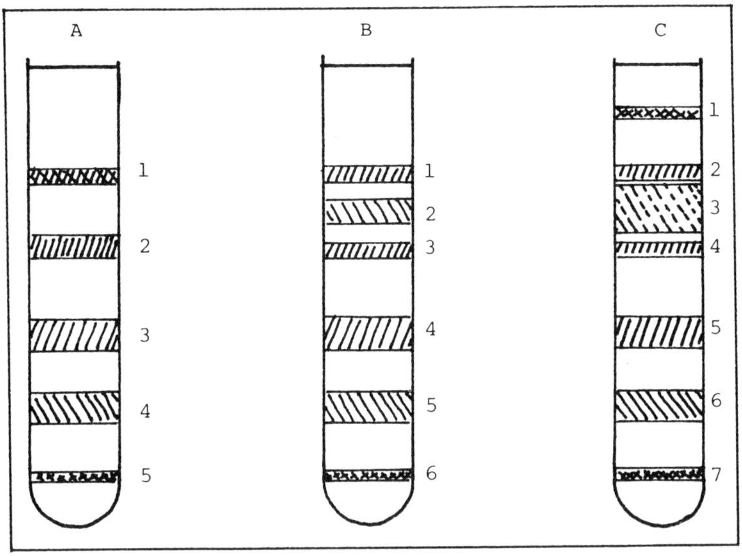

Figure 10. Banding patterns of smooth microsomes in iodinated density-gradient media. (A) 10–40% Nycodenz; (B) 10–40% metrizamide; (C) 10–40% metrizamide; microsomes homogenised with 10 additional strokes of the pestle.

5.2.2. *Fractionation of Smooth Microscomes*

The resolution of smooth microsomes in both metrizamide and Nycodenz depends to some extent on the severity of disruption of the membrane pellet prior to loading on top of the gradient. *Figure 10* shows the resolution of smooth microsomes, homogenised using 10 strokes of the pestle (see section 5.1 above). Five major bands are obtained in Nycodenz; six in metrizamide: the additional band in metrizamide occurs between bands 1 and 2 in Nycodenz. The enzyme profile of these bands is given in *Figures 11* and *12*. As in the case of the rough microsomes, the 5′-nucleotidase activity predominates in the lighter bands, although in metrizamide the spread of activity into the denser bands is more significant than in Nycodenz. In both gradients the galactosyl transferase is almost entirely confined to the least dense band. The NADPH-cytochrome reductase activity is broadly distributed in both gradients with a preponderence in the most rapidly sedimenting fractions; on the other hand the NADH-cytochrome c reductase distribution is significantly different in Nycodenz gradients while in metrizamide (not shown) the activities of both of the electron transport enzymes overlap entirely with only small differences in the relative amounts of the two enzymes in each band.

By doubling the number of strokes of the pestle during the resuspension of the smooth microsomal pellet, the banding pattern in the upper half of the gradient changes slightly. In metrizamide gradients seven bands are now apparent, as shown in *Figure 10*, and there are some small but significant changes in the pattern of enzyme distribution through the gradient (*Figure 13*). In spite of the tendency of the 5′-nucleotidase activity to be distributed rather similarly through most of

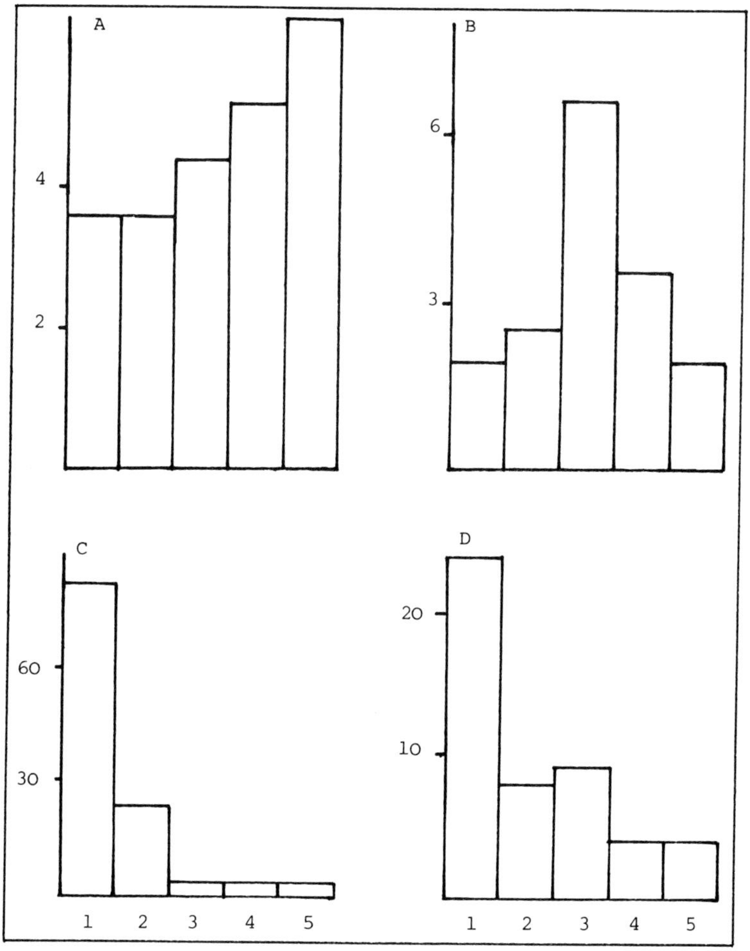

Figure 11. Subfractionation of smooth microsomes in 10–40% Nycodenz gradients: enzyme distribution. (A) NADPH-cytochrome c reductase; (B) NADH-cytochrome c reductase; (C) galactosyl transferase; (D) 5′-nucleotidase. All enzyme activities (ordinates) are given in μmoles substrate utilised/h/mg protein, except galactosyl transferase which is given in μg galactose transferred/h/mg protein.

the fractions, except the densest two (bands 6 and 7), the resolution from the NADPH-cytochrome c reductase is rather better. Like the 5′-nucleotidase, the galactosyl transferase is also distributed rather more widely in the gradient, but the clear maximum activity occurring in bands 2 and 3 distinguishes it from the 5′-nucleotidase. The NADH-cytochrome c reductase activity in this gradient overlaps the NADPH-cytochrome c reductase with only minor variations (*Figure 14a*). Only when the NADH cytochrome c reductase and NADPH-cytochrome c reductase activities are expressed as a ratio in each fraction is some significant dichotomy observed in that the relative enrichment of the NADH-cytochrome c reductase is highest in band 6 and lowest in band 2 (see *Figure 14a*).

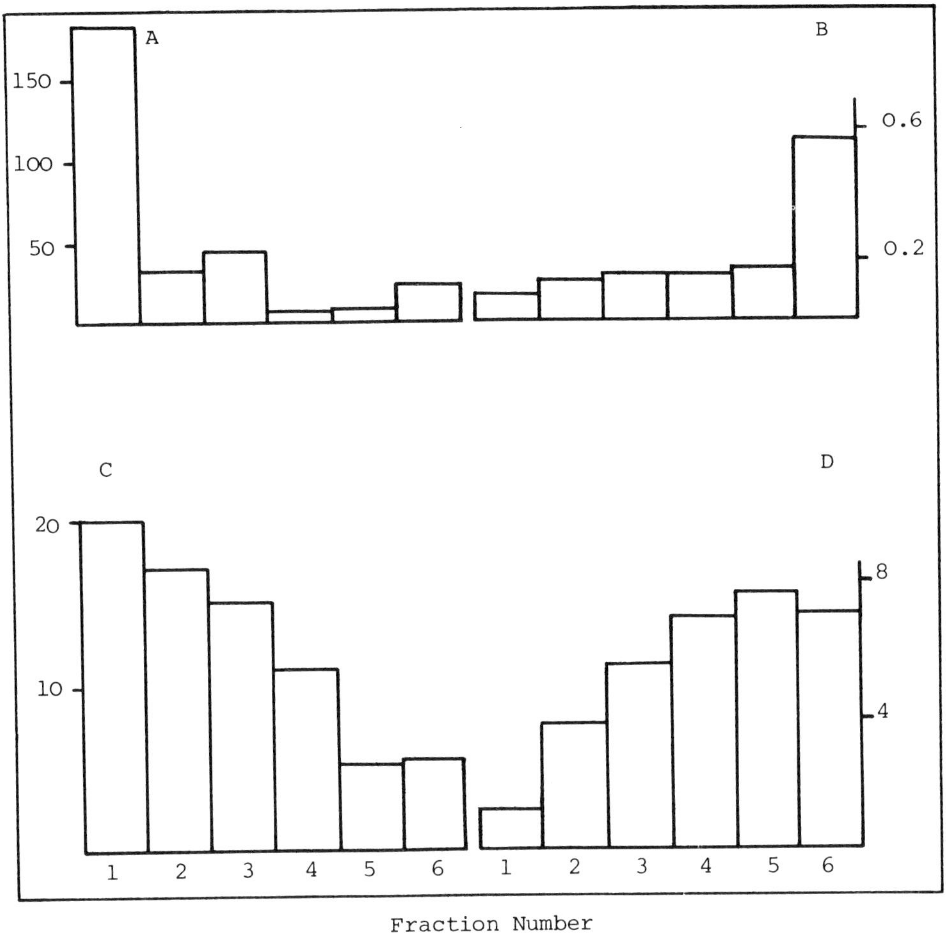

Figure 12. Subfractionation of smooth microsomes in 10–40% metrizamide gradients: enzyme distribution. (A) Galactosyl transferase; (B) glucose-6-phosphatase; (C) 5′-nucleotidase; (D) NADPH-cytochrome c reductase. For further details see legend to *Figure 11*.

Fractionating the total smooth microsomes on the basis of isopycnic density (using the same metrizamide gradient centrifuged at 100 000xg for 16 h) fails to improve this resolution although the details of the separation are changed in a significant way in as much as the least dense band is now relatively enriched in NADH-cytochrome c reductase compared to the other three bands (see *Figure 14b*).

From the existence of multiple fractions containing 5′-nucleotidase which exhibit varying amounts of galactosyl transferase activity (*Figure 13*) it is tempting to speculate that these may represent different parts of the cytoplasmic vesicle traffic to and from the surface membrane. In fraction 1, for example, the ratio of 5′-nucleotidase to galactosyl transferase is approximately eight times greater than in fraction 3. To investigate the possibility of resolving these fractions further, subfractionation in small-volume shallow gradients should be used, as described in the following section.

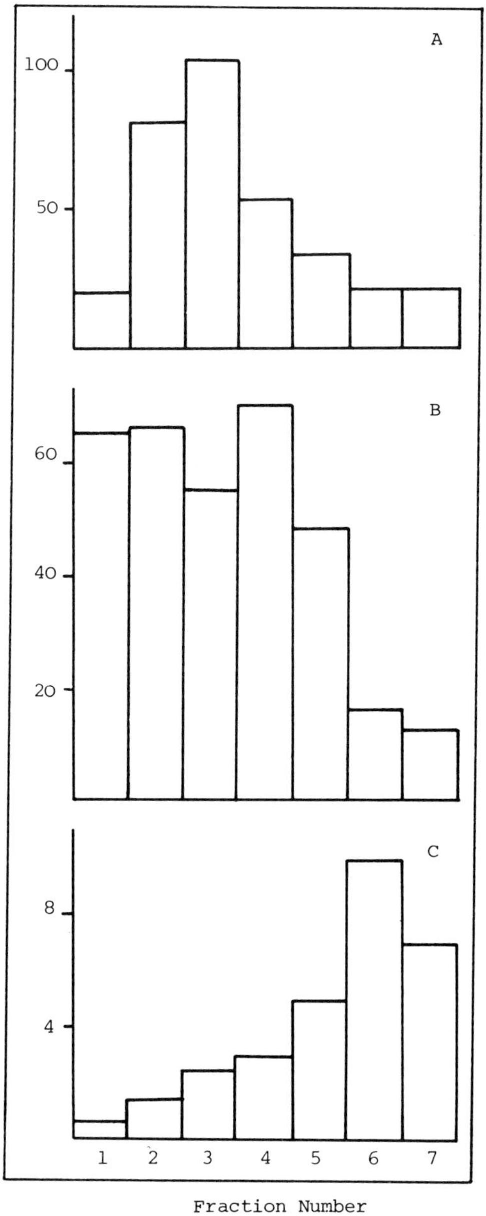

Figure 13. Subfractionation of smooth microsomes in 10–40% metrizamide gradients (microsomes homogenised with 20 strokes of the pestle): enzyme distribution. (A) Galactosyl transferase; (B) 5′-nucleotidase; (C) NADPH-cytochrome c reductase. For further details see legend to *Figure 11*.

5.2.3 *Subfractionation of Smooth Microsomes*

The bands are harvested from the first gradient and the concentration of metrizamide in each fraction determined by refractometry (11). If, for example, this

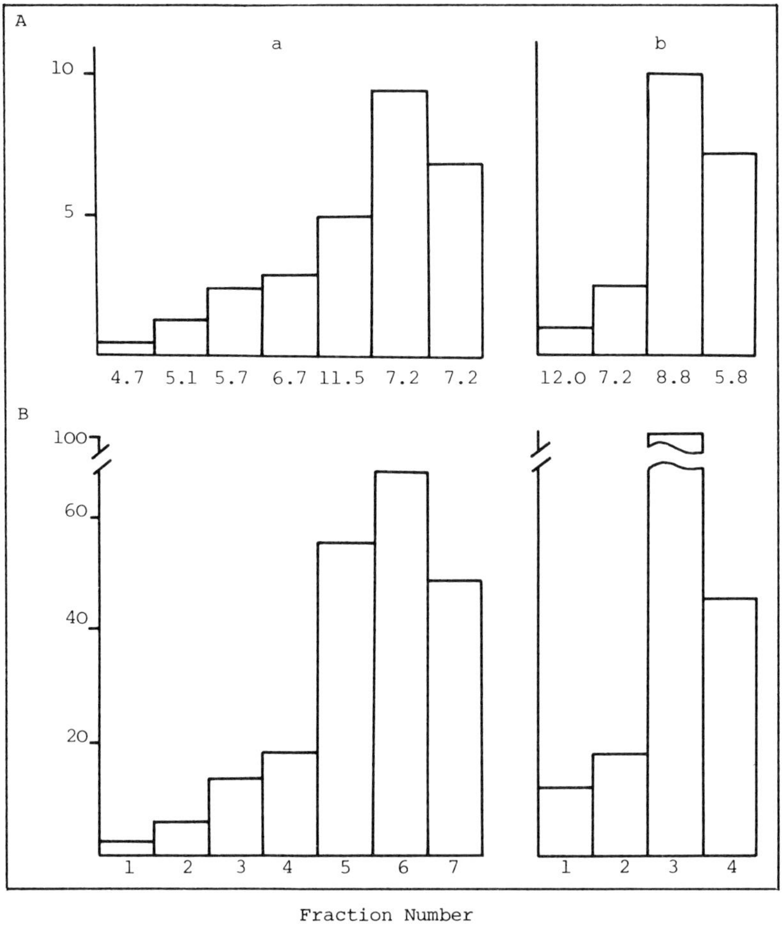

Figure 14. Subfractionation of smooth microsomes in 10–40% metrizamide gradients (microsomes homogenised with 20 strokes of the pestle) centrifuged at (a) 160 000xg for 1 h; (b) 100 000xg for 16 h: distribution of electron transport enzymes. (A) NADPH-cytochrome c reductase; (B) NADH-cytochrome c reductase, expressed as μmoles cytochrome c reduced/h/mg protein. Figures between upper and lower panels = NADH-cytochrome c reductase:NADPH-cytochrome c reductase ratio.

happens to be 24% and the volume is 0.7 ml, the percentage of metrizamide and volume are adjusted to 25% and 1.0 ml respectively and made part of a discontinuous gradient of 20%, 25%, 30% and 35% (all 1 ml) and centrifuged in an MSE 6 x 4.2 ml (Ti) swing-out rotor at 100 000xg for 12 h.

The details of this procedure for each band are shown in *Figure 15*. This procedure avoids the necessity for centrifugation and resuspension of the material from the first gradient and thus further comminution of the vesicles and possible loss of surface proteins is eliminated. The enzyme profiles of some selected fractions are shown in *Figure 16*. The subfractions from band 3 (bands 3-1 to 3-5) show a very clear separation of the major enzyme components 5′-nucleotidase and galactosyl transferase, whilst the subfractions from band 2 (bands 2-1 to 2-3)

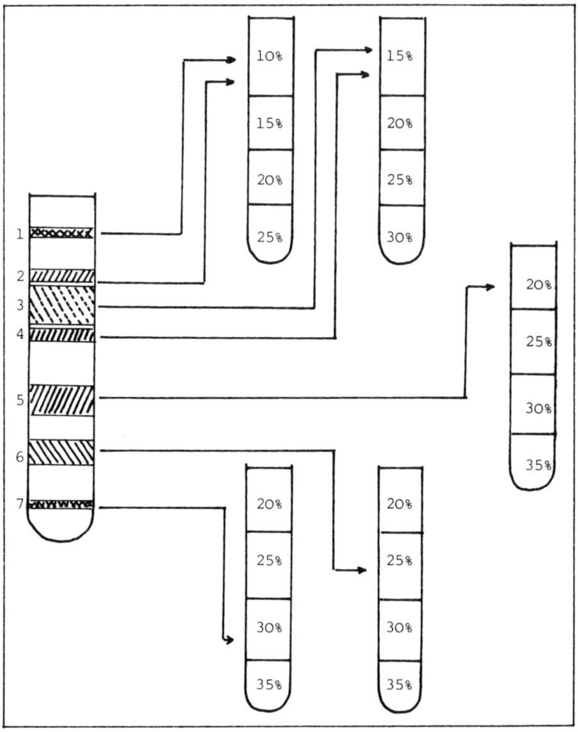

Figure 15. Subfractionation scheme for bands harvested from gradient C (see *Figure 10*).

demonstrate the existence of one population of vesicles which contains high levels of both of these enzymes. Also of considerable interest is the existence of two quite distinct fractions of Golgi (galactosyl transferase-containing) membranes which are of quite distinct densities, those from band 2 (bands 2-1 and 2-2) have densities of around 1.080 g/ml whilst those from band 3 (bands 3-3 and 3-4) have densities of around 1.130 g/ml. The major NADPH-cytochrome c reductase containing fraction (band 6-3) contains only very low levels of 5′-nucleotidase and galactosyl transferase and represents a more than 60-fold purification compared with the homogenate. The electron-transport enzymes in the subfractions from bands 5 and 6 still tend to co-purify (*Figure 17*). On the other hand, when the activity ratios are compared (see *Figure 17*) in the subfraction from band 5 and band 6, it is clear that those from band 6 (and band 6-3 in particular) are significantly enriched in the NADPH-cytochrome c reductase enzyme.

5.3. Important Practical Points

(i) The homogenisation of the microsome pellets prior to fractionation is critical in producing reproducible banding patterns in the subsequent gradients. Specific conditions are almost certain to vary from laboratory to laboratory, but every attempt should be made to standardise (a) the concentration of microsomal

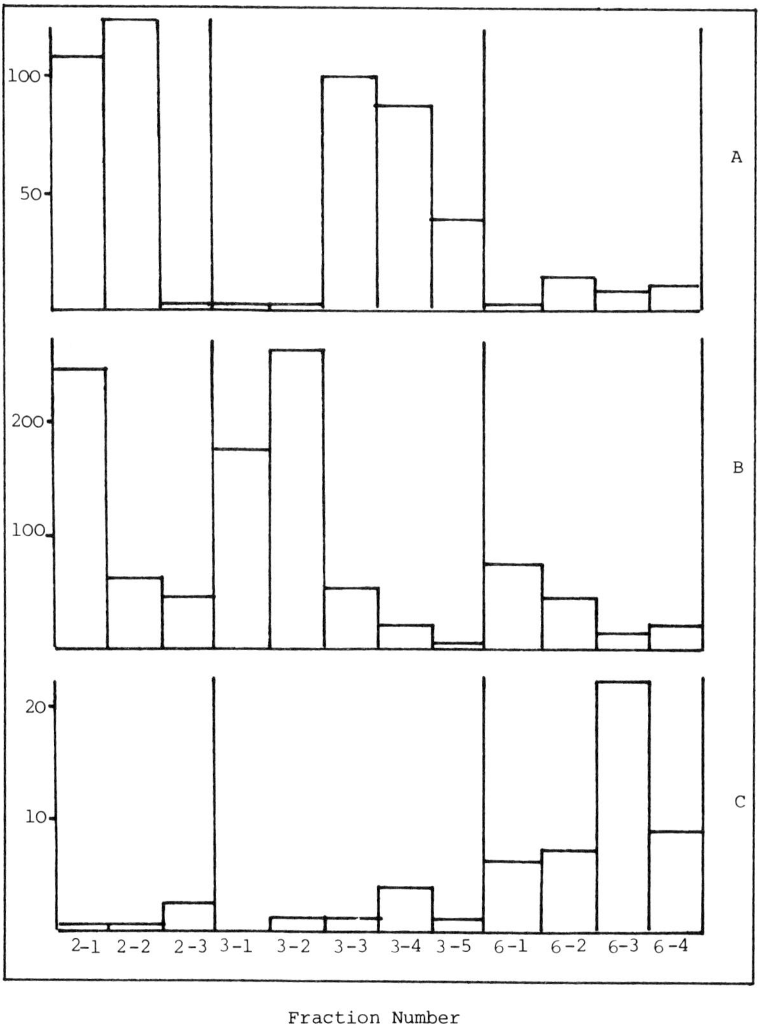

Figure 16. Subfractionation of bands 3, 4 and 7 from gradient C (see *Figure 10*): enzyme distribution. (A) Galactosyl transferase; (B) 5′-nucleotidase; (C) NADPH-cytochrome c reductase. For further details see legend to *Figure 11*.

material; (b) the volume of suspension; (c) the homogenisation conditions especially the number of strokes of the pestle, speed of rotation of the pestle and the clearance of the homogeniser.

(ii) Metrizamide may interfere with galactosyl transferase assays since the molecule includes glucosamine linked to the triiodinated benzene ring (in the assay glucosamine is used as an acceptor of galactose), although we have no indication that significant interference occurs in sensitive assays, in which the concentration of metrizamide is less than 2% in the assay medium. To avoid any possible effects of the gradient medium on any enzyme assay it is advisable to transfer the membrane sample into 0.25 M sucrose.

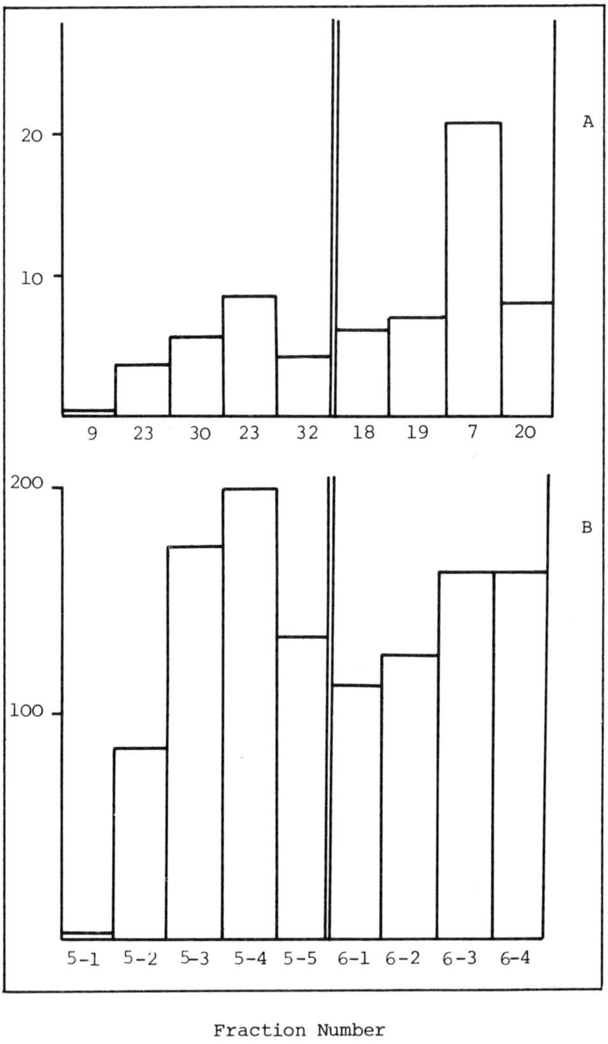

Figure 17. Subfractionation of bands 6 and 7 from gradient C (see *Figure 10*): distribution of electron transport enzymes. For further details see legend to *Figure 14*.

(iii) Fractionations designed to isolate a specific type of enzyme enriched vesicle may require alterations to the experimental protocols used in these studies. Parameters which may affect the fractionation are: time and speed of centrifugation, concentration of divalent cations in the gradient and lipid/protein ratio of the vesicles. Some vesicles may only be resolvable after washing in hypotonic buffer to remove trapped soluble protein and/or protein adsorbed on to the surface of the vesicle.

5.4. Conclusions

Gradients of metrizamide and Nycodenz can effect subfractionation of both the rough and smooth microsomes from rat liver far more efficiently than those of sucrose inasmuch as discrete, well-separated bands of material are obtained. By using a combination of rate sedimentation and isopycnic separations in metrizamide, the existence of enzyme heterogeneity within the endoplasmic reticulum can be examined more clearly.

The ability of metrizamide to produce quite distinct subfractions of the Golgi which have a unique enzyme composition has not been observed with sucrose gradients and the potential of iodinated density-gradient media for dissecting the complex vesicle traffic within the hepatocyte is considerable.

Data suggest that the separations obtainable with metrizamide and Nycodenz, though similar are not identical, and in future a judicious combination of the two media may optimise the fractionations of membranes.

6. FRACTIONATIONS OF TISSUE CULTURE CELL MEMBRANES

It is not possible to make any generalised recommendations regarding the homogenisation and fractionation of tissue culture cells because of the great variability in the susceptibility of different lines of cultured cells and even of the same cell line cultured for different lengths of time, to the shearing forces of Dounce and Potter-Elvehjem homogenisation (21). However, it is important to point out that that metrizamide gradients are useful in subfractionation of both plasma membrane and endoplasmic reticulum vesicles from these cells.

It is only necessary to include a few suggestions regarding homogenisation. The aims of homogenisation should be to achieve at least 90% cell rupture, with minimal nuclear damage as monitored by phase-contrast microscopy, using fewer than 20 strokes of the pestle of a Dounce or Potter-Elvehjem homogeniser. To avoid the requirement for prolonged homogenisation most tissue culture cells require osmotic swelling. Hypotonic media which have been used include 1 mM $NaHCO_3$ (pH 8.0), 20 mM Tris-HCl (pH 7.4–8.0), 10 mM imidazole-HCl (pH 8.0). To protect the nuclei from breaking, the media can be supplemented with Mg^{2+} (0.2–1.0 mM), Ca^{2+} (0.2–1.0 mM) or K^+ (1–10 mM), either singly or in combination. Protection of internal organelles is also afforded by the released cytosolic proteins and the cell/homogenisation medium volume ratio should be as large as possible to maximise this protection. Again this varies from cell to cell but as a general guideline 10^9 cells/20 ml may be used as a starting point. For a more thorough examination of this problem see ref. 21.

To characterise the products of homogenisation, the homogenate, which should be made 0.25 M with respect to sucrose immediately after homogenisation, is subjected to differential centrifugation. A suitable sequence of centrifugation steps is:

(i) 1000xg for 10 min;
(ii) 2000xg for 10 min;
(iii) 5000xg for 10 min;
(iv) 10 000xg for 20 min;
(v) 20 000xg for 20 min;
(vi) 100 000xg for 60 min.

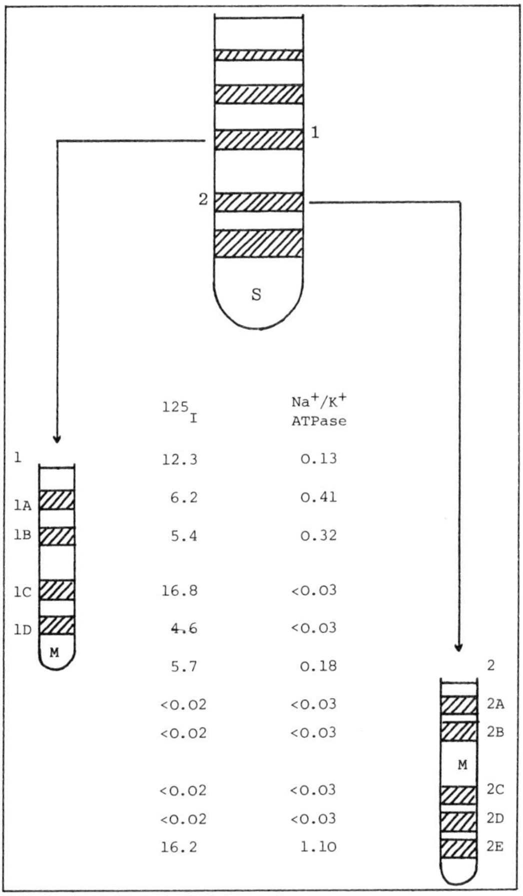

Figure 18. Subfractionation of material pelleted by centrifugation at 10 000xg for 20 min from ^{125}I-lactoperoxidase-labelled Lettree cells in sucrose (S) and metrizamide (M) gradients. ^{125}I activity is expressed as c.p.m./μg protein; Na^+/K^+-ATPase activity is expressed as μmoles ATP hydrolysed/h/mg protein. The experimental method is as described in the text.

Each pellet should be analysed for Na^+/K^+-ATPase, NADPH-cytochrome c reductase and succinate-cytochrome c reductase to determine the fractionation patterns of plasma membrane, endoplasmic reticulum and mitochondria, respectively. Ideally fragments of the plasma membrane should sediment at considerably lower centrifugation speeds (in steps (iii) or (iv)) than do the endoplasmic reticulum vesicles (in steps (v) or (vi)) since it is the separation of plasma membrane and smooth endoplasmic reticulum vesicles which is the most problematical aspects of tissue culture cell membrane fractionation. Use of the gentlest shearing forces during homogenisation (i.e. the lowest number of strokes of the pestle and largest pestle clearance) will maximise the difference in plasma membrane and endoplasmic reticulum sedimentation.

The fractionation of step (iv) and step (vi) pellets from ^{125}I-lactoperoxidase-labelled Lettree cells will be given as an example of the improved resolving power

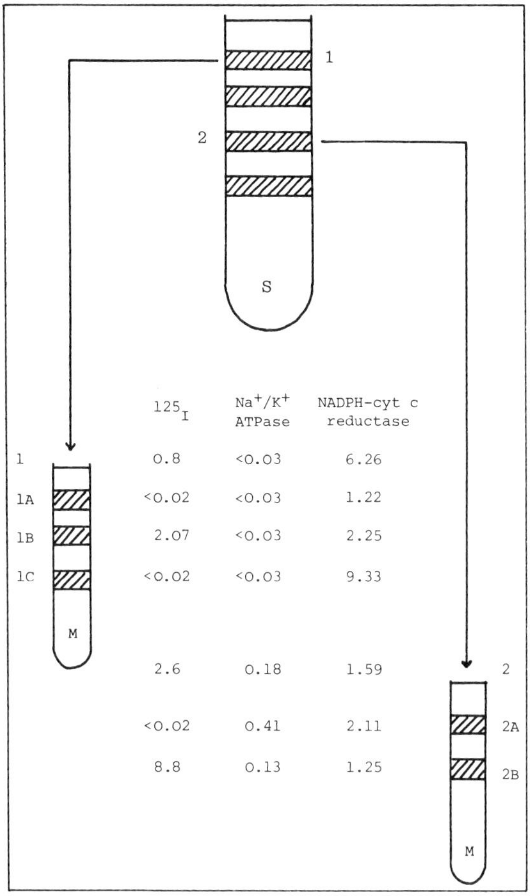

Figure 19. Subfractionation of material pelleted by centrifugation at 100 000xg for 60 min from ^{125}I-lactoperoxidase-labelled Lettree cells in sucrose (S) and metrizamide (M) gradients. ^{125}I activity is expressed as c.p.m./μg protein; Na^+/K^+-ATPase activity is expressed as μmoles ATP hydrolysed/h/mg protein; NADPH-cytochrome c reductase activity is expressed as μmoles cytochrome c reduced/h/mg protein. The experimental method is as described in the text.

obtainable using metrizamide gradients (16). The pellets should be suspended in 4 ml of 30% (w/w) sucrose and made part of a discontinuous gradient prepared by layering 4 ml each of 25%, 35%, 40%, 45% and 50% sucrose in an MSE 3 x 25 ml swing-out rotor. All the solutions should contain 5 mM Tris-HCl (pH 8.0). After centrifugation at 90 000xg for 16 h, harvest the bands, dilute each with three volumes of 5 mM Tris-HCl (pH 8.0) and centrifuge at 100 000xg for 60 min. Resuspend the pellets in 0.7 ml 10% (w/v) metrizamide, 5 mM Tris-HCl (pH 8.0). Layer each sample onto a gradient prepared by layering 0.7 ml each of 20%, 25%, 30%, 35% and 40% metrizamide; the gradients are centrifuged at 300 000xg for 2 h. Examples of the additional resolution in metrizamide are shown in *Figures 18* and *19*. *Figure 18* shows that the step (iv) pellet fractionates into five bands in the sucrose gradient. Two of these, upon further fractionation in metrizamide gradients, produce subfractions which indicate the possible heterogeneity which exists even at the surface of a tissue culture cell, for example bands 1C and 2E both contain high levels of ^{125}I but only band 1C contains Na^+/K^+-ATPase. The original bands from the sucrose gradient (bands 1 and 2) give very little indication of such a dichotomy in these activities. *Figure 19* demonstrates four bands in the sucrose gradient and a significant improvement in the purity of an endoplasmic reticulum fraction (band 1C) is obtained in metrizamide such that no surface ^{125}I is detectable. Subfraction 3A is also interesting as it is apparently an Na^+/K^+-ATPase fraction not derived from the surface.

7. ACKNOWLEDGEMENTS

The authors would like to express their gratitude to the Cell Surface Research Fund and the Medical Research Council for financial support.

8. REFERENCES

1. Wisher,M.H. and Evans,W.H. (1975) *Biochem. J.,* **146**, 375.
2. Eicholz,A. (1967) *Biochim. Biophys. Acta,* **135**, 475.
3. Colas,B. and Maroux,S. (1980) *Biochim. Biophys. Acta,* **600**, 406.
4. Fujita,M., Ohta,H. and Uezato,T. (1981) *Biochem. J.,* **196**, 669.
5. Kenny,A.J. and Booth,A.G. (1970) *Biochem. Soc. Trans.,* **4**, 1011.
6. Palade,G. (1975) *Science,* **189**, 347.
7. Evans,W.H. (1980) *Biochim. Biophys. Acta,* **604**, 27.
8. Neville,D.M. (1976) in *Biochemical Analysis of Membranes* (Maddy,A.H. ed.) Chapman and Hall, London, p. 27.
9. Regoeczi,E., Chindema,P.A., Debanne,T. and Charlwood,P.A. (1982) *Proc. Natl. Acad. Sci. (USA),* **79**, 2226.
10. Debanne,M.T., Evans,W.H., Flint,N. and Regoeczi,E. (1982) *Nature,* **298**, 398.
11. Rickwood,D. and Birnie,G.D. (1975) *FEBS Lett.,* **50**, 102.
12. Rickwood,D., Ford,T. and Graham,J.M. (1982) *Anal. Biochem.,* **123**, 33.
13. Wattiaux,R., Wattiaux-de Coninck,S. and Ronveaux-Dupal,M.F. (1978) *J. Cell Biol.,* **78**, 349.
14. Wattiaux,R., Wattiaux-de Coninck,S. and Vandenberghe,A. (1982) *Biol. Cell.,* **45**, 474.
15. Aas,M. (1973) Proc. 9th Internat. Congr. Biochem., Stockholm, p. 31.
16. Graham,J.M. and Coffey,K.H.M. (1979) *Biochem. J.,* **182**, 173.
17. Hubscher,G., West,G.R. and Brindley,D.N. (1965) *Biochem. J.,* **97**, 629.
18. Bergstrand,A. and Dallner,G. (1969) *Anal. Biochem.,* **29**, 351.
19. Eriksson,L.C. (1973) *Acta Pathol. Microbiol. Scand. A.,* suppl. 239.
20. Svennson,H., Dallner,G. and Ernster,L. (1972) *Biochim. Biophys. Acta,* **274**, 447.
21. Graham,J.M. (1982) in *Cancer Cell Organelles* (Reid,E., Cook,G.M.W. and Moore,D.J. eds) Ellis Horwood, Chichester, England, p. 343.

CHAPTER 6

Separation of Cell Organelles

R. Wattiaux and S. Wattiaux-De Coninck

1. INTRODUCTION

When a subcellular structure is centrifuged in a density gradient, its size, shape and density can change during the centrifugation. It is possible to take advantage of these changes that modify the sedimentation coefficient of the particles to analyse some properties of the particles and to isolate them. However, centrifugation is most frequently performed in a density gradient so that the particles reach the density zone exhibiting a similar density to their own. Having reached this density zone, particles do not migrate further since their speed of sedimentation is equal to zero. Such centrifugation is generally called isopycnic centrifugation; in these conditions, only the density of the particle determines its distribution in the gradient. It is this type of centrifugation that will be considered in this chapter. The first part of this chapter discusses the behaviour of a particle subjected to isopycnic centrifugation in a density gradient of sucrose or metrizamide. In the second part the problem of the purification of lysosomes, peroxisomes and mitochondria by gradient centrifugation is discussed and the usefulness of metrizamide for such separations is described, together with a short comment on the use of Nycodenz gradients.

2. BEHAVIOUR OF A PARTICLE IN A DENSITY GRADIENT

2.1. Factors that Influence the Density of a Subcellular Organelle

Each subcellular organelle is surrounded by a membrane. The permeability of this membrane to the solute used to make the density gradient predominantly affects the density changes of the particle when it sediments through the gradient. This can be very simply illustrated. If one takes, as an example, a particle whose membrane is freely permeable to the external solute then, during its migration through the gradient, the density of the particle will continuously increase owing to the penetration of more and more solute. Now, let us suppose that the particle membrane is impermeable to the external solute. In this case, the density of the particle can change as a result of a loss or gain of water caused by continuous changes in the osmotic environment.

Obviously the situation is more complicated if interactions other than the one resulting from the permeability of the particle membrane take place between the particle and the solute or if the permeability of the membrane is affected by the centrifugation procedure.

2.2. **Models for Mitochondria, Lysosomes and Peroxisomes**

In the case of sucrose solutions the influence of the composition of the medium on the density of mitochondria, lysosomes and peroxisomes has been thoroughly investigated by de Duve's group (1,2). These authors have proposed that mitochondria and lysosomes consist of three spaces: one is freely accessible to sucrose (the sucrose space); it is thus always filled with a sucrose solution of a concentration equal to that of the outside medium; a second space is characterised by the amount of osmotic water which the particle contains; the third is made up of all the solid components of the particle (the matrix space). Peroxisomes do not have an osmotic space.

The volume, mass and density of these spaces and, therefore, of the total particle, are related to the sucrose concentration of the medium by relationships involving parameters characteristic of each space. For example, the density of the particles is given by the following equation which is valid for mitochondria and lysosomes

$$\varrho_p = \varrho_w \frac{\varrho_d\alpha + (\varrho_d + \varrho_m\beta)m}{\varrho_d\alpha + \varrho_w(1 + \beta)m}$$

where ϱ_w = density of the solvent in g/ml

ϱ_d = density of the granule matrix which includes all the solid components of the particle, in g/ml

α = particle relative content of osmotically active solutes in milliosmoles/gram of hydrated matrix

ϱ_m = medium density in g/ml

β = relative volume of the particle space freely accessible to sucrose, in ml/ml hydrated matrix

m = molality of the sucrose solution in millimoles/gram of solvent

In contrast the density of peroxisomes, which have no osmotic space, changes according to the following relationship

$$\varrho_p = \frac{\varrho_d + \varrho_m\beta}{1 + \beta}$$

2.3. **Determination of Parameters α, β and ϱ_d**

Beaufay and Berthet (1) have determined the most probable values of α, β and ϱ_d by measuring the equilibrium densities of the particles in glycogen gradients

Table 1. Parameters α, β and ϱ_d for Rat-Liver Mitochondria, Lysosomes and Peroxisomes. α is given in milliosmoles/gram of hydrated matrix, β in ml/ml of hydrated matrix and ϱ_d in g/ml. Data are those of Beaufay and Berthet (1), calculated by taking into account the behaviour of cytochrome c oxidase, acid phosphatase and catalase, reference enzymes for mitochondria, lysosomes and peroxisomes, respectively.

Parameters	Mitochondria	Lysosomes	Peroxisomes
α	0.110	0.102	0
β	0.760	1.05	2.55
ϱ_d	1.200	1.216	1.230

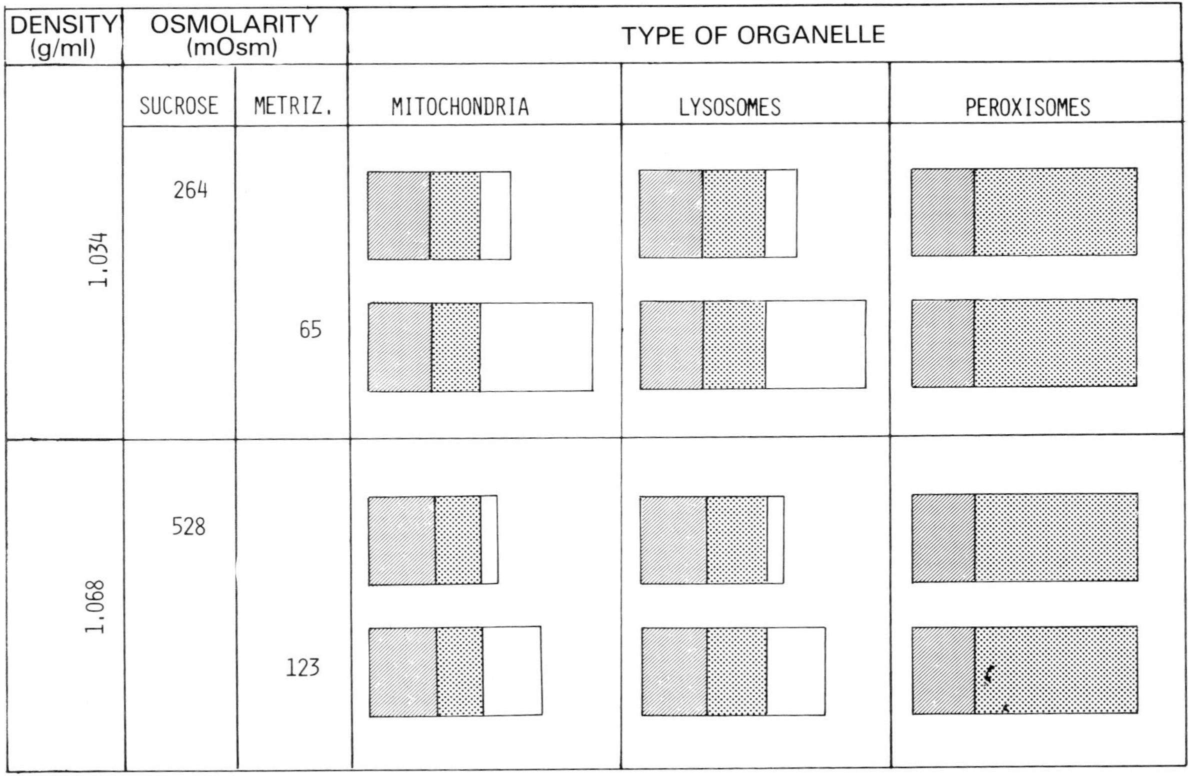

Figure 1. Relative volumes of matrix, sucrose and osmotic space for rat-liver mitochondria, lysosomes and peroxisomes in sucrose and metrizamide solutions with densities of 1.034 g/ml and 1.068 g/ml. The values have been calculated by using the parameters α, β and ϱ_d given in *Table 1*.

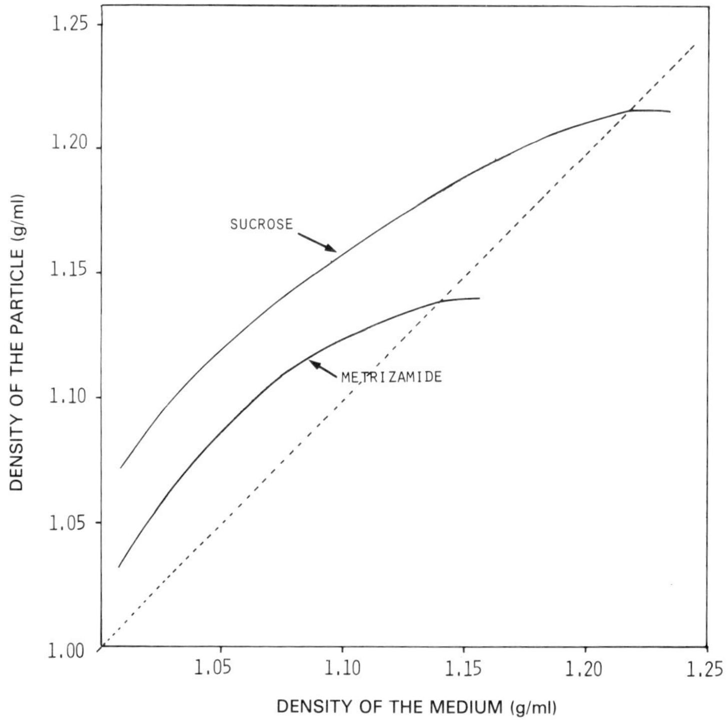

Figure 2. Theoretical changes of rat-liver mitochondria as a function of the density of sucrose or metrizamide solutions. The curves intersect the diagonal at the equilibrium density of the particle in a sucrose or in a metrizamide gradient.

containing various concentrations of sucrose. This method is based on the fact that glycogen, the substance used to make the gradient, cannot cross the membrane of organelles and, therefore, cannot influence their density. In these conditions, the density of the particles is only a function of the sucrose solution chosen as the solvent. The values are given in *Table 1.*

Knowing the parameters α, β and ϱ_d, it is possible to predict the behaviour of mitochondria, lysosomes and peroxisomes in a sucrose gradient, as will be seen later, or in a gradient made of a macromolecular or particulate gradient medium with a sucrose solution as the solvent. For example, such a calculation could be used to predict the behaviour of these organelles in a Percoll gradient prepared in a sucrose medium.

2.4. Theoretical Behaviour of Mitochondria, Lysosomes and Peroxisomes in Metrizamide

When a solute exhibits some similarity to sucrose with respect to its properties, it seems justified to make use of the parameters and relationships described in Section 2.3 to predict the behaviour of mitochondria, lysosomes and peroxisomes in aqueous solutions of this solute. This is what has been done in the

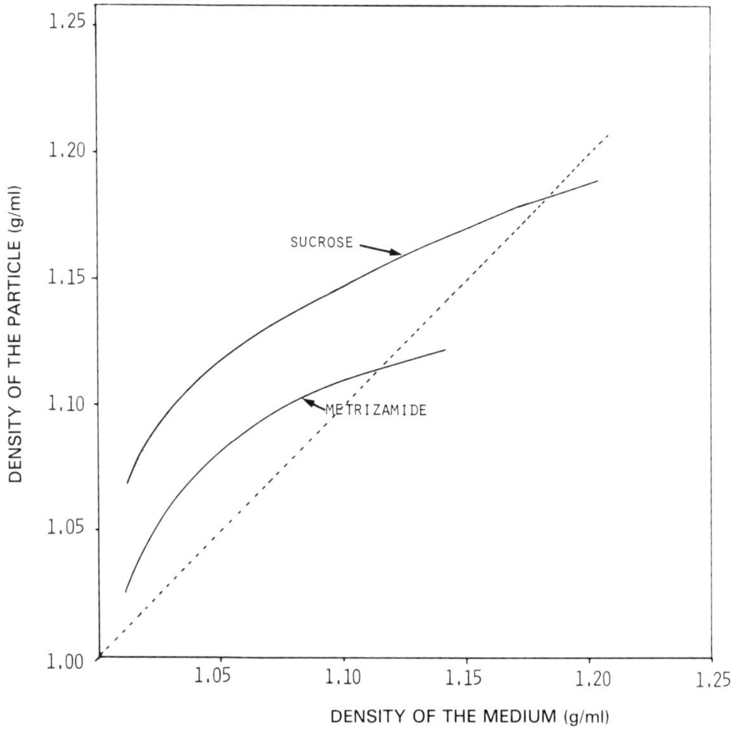

Figure 3. Theoretical changes of rat-liver lysosomes as a function of the density of sucrose or metrizamide solutions. The curves intersect the diagonal at the equilibrium density of the particle in a sucrose or in a metrizamide gradient.

case of metrizamide. *Figure 1* shows how the volumes of the spaces would be expected to be distributed in mitochondria and lysosomes in solutions of metrizamide and sucrose of equal densities. A net difference becomes apparent for the volumes of the osmotic space which are markedly larger in metrizamide solutions than in sucrose solutions, the reason being that, at equal densities, metrizamide solutions exhibit a lower osmolarity. As a result, the density of the particle is lower in metrizamide than in sucrose. In the case of peroxisomes, no such difference occurs since these granules are devoid of an osmotic space. These considerations allow one to predict the possible equilibrium density of these particles in sucrose as well as in metrizamide gradients.

2.5. **Equilibrium Density in Sucrose and Metrizamide Gradients**

Figures 2, 3 and *4* represent the changes in the densities of mitochondria, lysosomes and peroxisomes as a function of the density of the medium, the solute being sucrose or metrizamide. In agreement with theoretical considerations, the densities of mitochondria and of lysosomes are always lower in metrizamide solutions than in sucrose solutions of equal density. In contrast, the curves are superimposable in the case of peroxisomes. It is possible to determine the possible equilibrium density in gradients made with these solutes

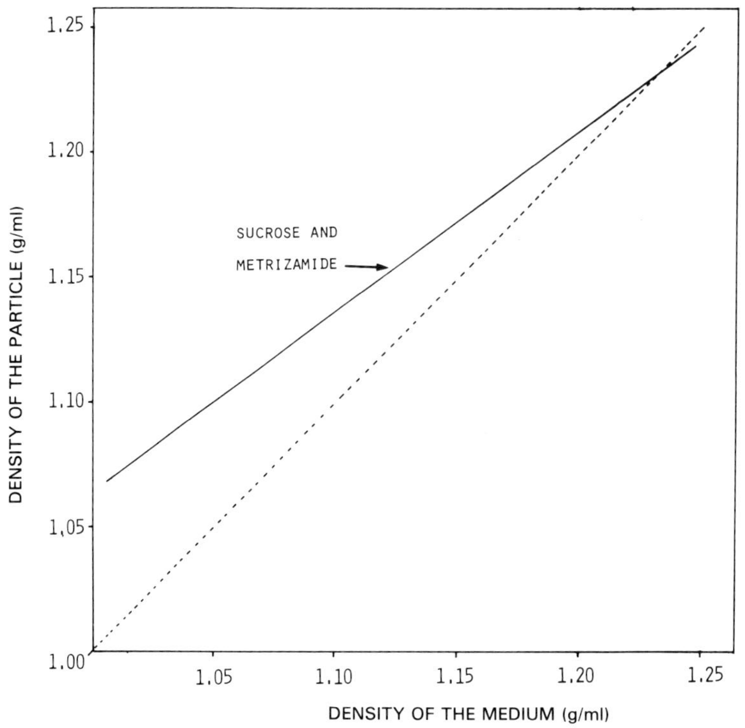

Figure 4. Theoretical changes of rat-liver peroxisomes as a function of the density of sucrose or metrizamide solutions. The line intersects the diagonal at the equilibrium density of the particle in a sucrose or in a metrizamide gradient.

by drawing a diagonal. The intersection of the diagonal with the curve gives the value of the equilibrium density of the particle in a density gradient since this point corresponds to the density of the medium which is equal to the density of the particle. Thus it can be seen that the equilibrium densities of mitochondria and lysosomes are distinctly lower in metrizamide gradients; obviously, in the case of peroxisomes the equilibrium densities are the same in both media.

2.6. **Experimental Results**

The experimental results are in good agreement with the theoretical predictions. Indeed, after isopycnic centrifugation, lysosomes and mitochondria band at a lower density in metrizamide gradients than they do in sucrose gradients; peroxisomes, however, band at the same density in both kinds of gradient. *Figures 5* and *6* illustrate the distribution patterns of β-galactosidase (a marker enzyme for lysosomes) and catalase (a peroxisome marker enzyme) after isopycnic centrifugation in a sucrose gradient and in a metrizamide gradient. As ascertained by the enzyme distributions, the median equilibrium den-

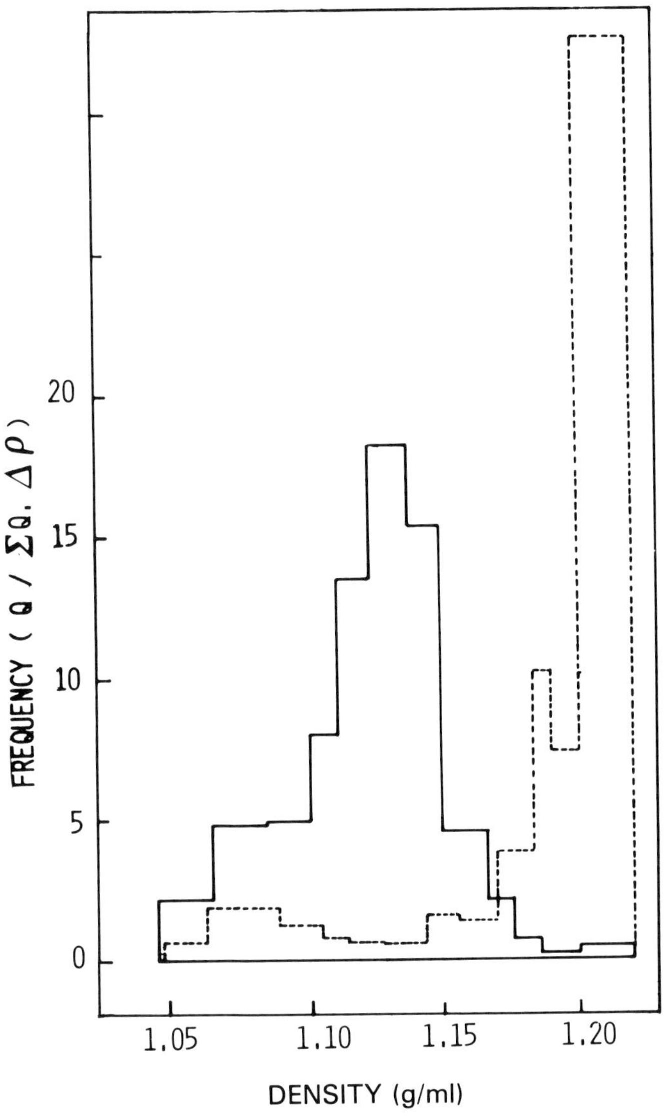

Figure 5. Distribution pattern of β-galactosidase after isopycnic centrifugation in a sucrose (------) or in a metrizamide (——) gradient of a rat-liver mitochondrial fraction (ML fraction of de Duve *et al.* ref. 3). The granules suspended in 0.25 M sucrose were layered onto the top of the gradient. The gradients were centrifuged at 110 000xg for 150 min at 5°C (Beckman rotor SW 65). Ordinate: average frequency of the components for each fraction $Q/\Sigma Q \cdot \Delta\varrho$ where Q represents the activity found in the fraction, ΣQ the total recovered activity and $\Delta\varrho$ the increments of density from top to bottom of the fraction.

sity of lysosomes is about 1.22 g/ml in sucrose gradients and 1.13 g/ml in metrizamide gradients. The density of peroxisomes is greater than 1.22 g/ml in both types of gradient.

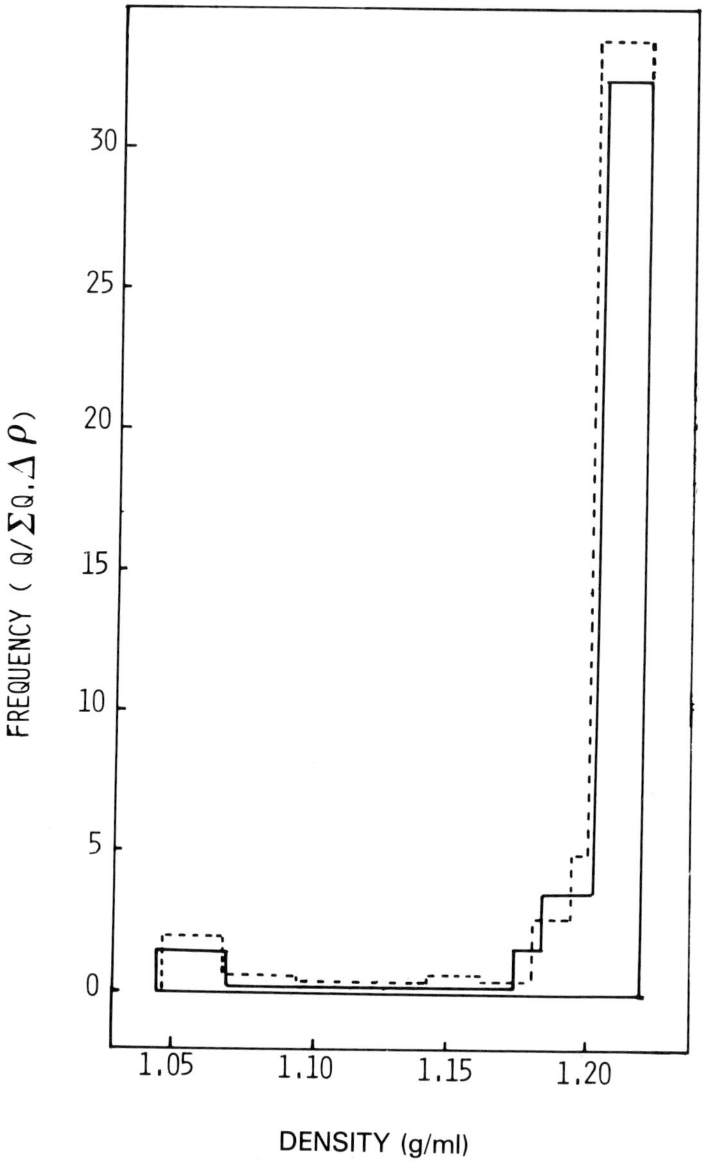

Figure 6. Distribution pattern of catalase after isopycnic centrifugation in a sucrose (------) or in a metrizamide (——) gradient of a rat-liver mitochondrial fraction (ML fraction of de Duve *et al.* ref. 3). The conditions of centrifugation and the mode of representation are the same as in *Figure 5.*

3. ISOLATION OF LYSOSOMES AND PEROXISOMES IN METRIZAMIDE GRADIENTS

3.1. Analytical Results

It is obvious that sucrose gradients are not suitable for separating lysosomes

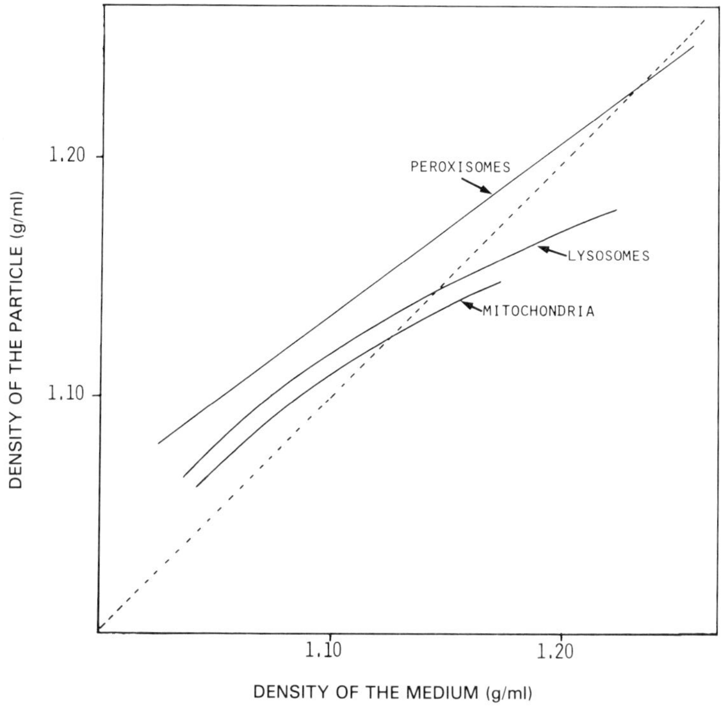

Figure 7. Theoretical changes of rat-liver mitochondria, lysosomes and peroxisomes as a function of the density of metrizamide solution. The lines intersect the diagonal at the equilibrium density of the particles in a metrizamide gradient.

from peroxisomes and mitochondria. Indeed the median equilibrium densities of these particles are too close to those of mitochondria and peroxisomes; moreover the situation is complicated by the fact that the organelle population, particularly that of lysosomes, is heterogeneous, as illustrated by the spread of the distribution curves of the lysosomal enzymes. On the other hand, as shown by *Figure 7*, it may be expected that a good separation of lysosomes and mitochondria from peroxisomes could be achieved by isopycnic centrifugation in a metrizamide gradient. *Figure 8* illustrates the distribution of β-galactosidase (lysosomes), cytochrome oxidase (mitochondria) and catalase (peroxisomes) after isopycnic centrifugation of a light mitochondrial fraction (L) of de Duve *et al.* (3) in a metrizamide gradient. Obviously, as shown by the distribution of catalase, peroxisomes equilibrate in a region clearly distinct from the region where mitochondria and lysosomes are found.

The separation of lysosomes from mitochondria would appear to be difficult to carry out using a metrizamide gradient; indeed both particles have an osmotic space and their density is similarly affected by metrizamide. Nevertheless, it is possible to obtain an excellent purification of lysosomes in a metrizamide gradient by taking advantage of the fact that the density of mitochondria can be increased if the particles are subjected to a sufficiently high hydrostatic pressure. As shown previously (4,5), this phenomenon probably

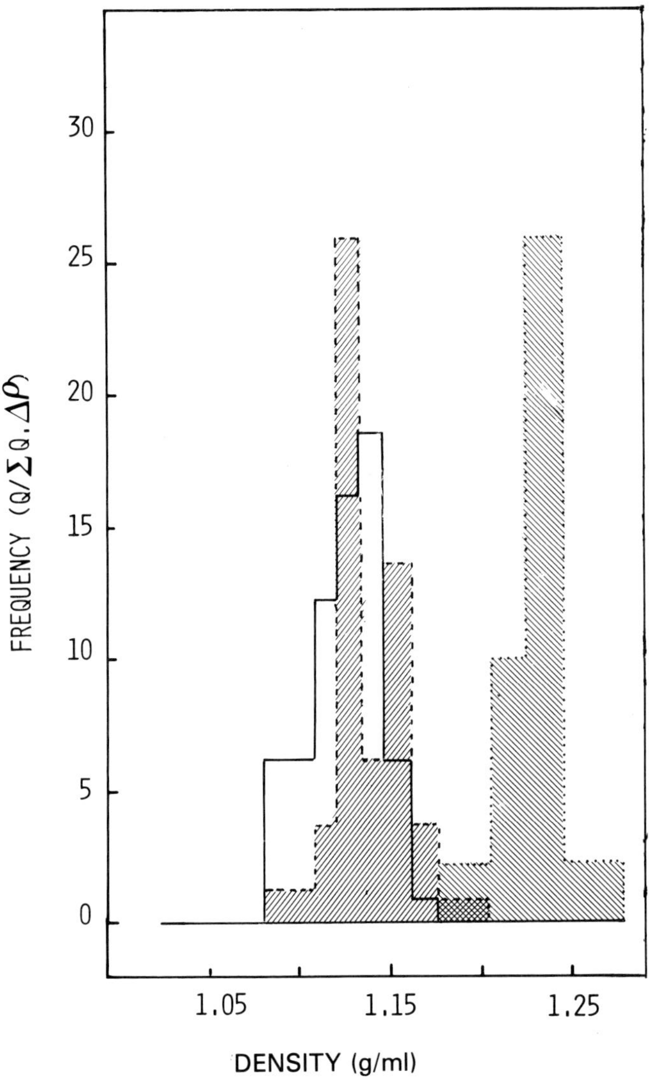

Figure 8. Distribution pattern of β-galactosidase □, cytochrome c oxidase ▨ and catalase ▧ after isopycnic centrifugation of rat-liver light mitochondrial fraction (L fraction of de Duve *et al.* ref. 3) in a metrizamide gradient. The granules suspended in 0.25 M sucrose were initially layered at the top of the gradient. The conditions of centrifugation and the mode of representation are the same as in *Figure 5*.

results from an effect of hydrostatic pressure on the inner mitochondrial membrane which then becomes permeable to sucrose. Lysosomes are markedly less affected by hydrostatic pressure. The hydrostatic pressure that the organelles are subjected to during centrifugation will be higher if they are layered at the bottom of the gradient, under the liquid column, before centrifugation. Under these conditions, the distribution curve of cytochrome oxidase is shifted

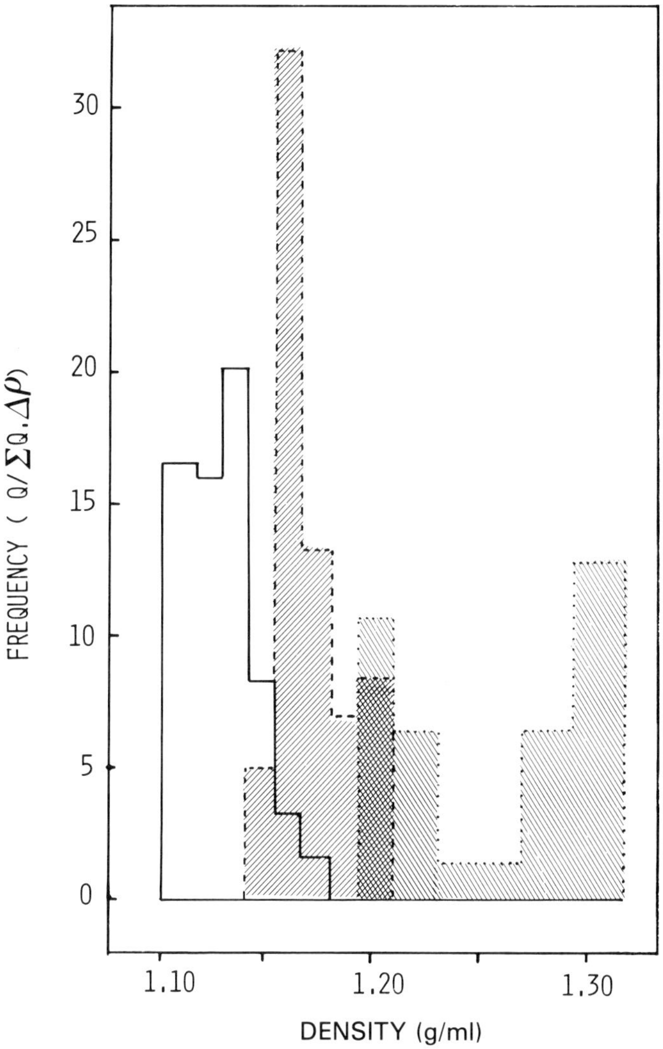

Figure 9. Distribution patterns of β-galactosidase, cytochrome c oxidase and catalase after isopycnic centrifugation of a rat-liver light mitochondrial fraction (L fraction of de Duve *et al.* ref. 3) in a metrizamide gradient. The granules suspended in 57% metrizamide were layered below the gradient. The conditions of centrifugation and the mode of representation are the same as in *Figure 5*.

towards higher density regions, and does not overlap the distribution curve of lysosomal hydrolases *(Figure 9)*.

3.2. Isolation Procedures

On the basis of the observations described in the previous section a procedure has been devised for purifying lysosomes and peroxisomes by centrifugation in a discontinuous metrizamide density gradient.

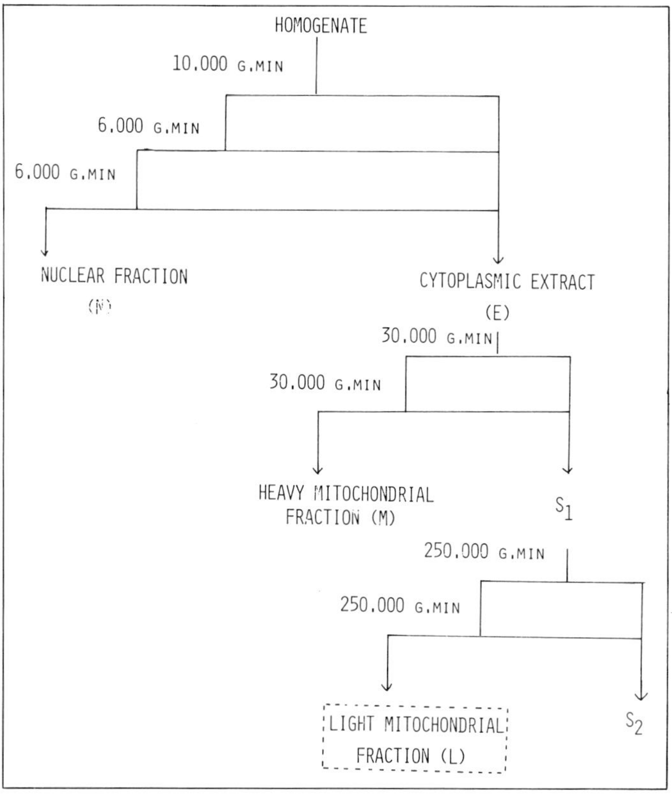

Figure 10. Preparation of L fraction according to de Duve *et al.* (3). The quantity g.min is a composite unit which corresponds to $\int_t^o g_{av} dt$, where t = the time of centrifugation; g_{av} = the average field prevailing at a distance $R_{av} = 0.5\ (R_{max} + R_{min})$ in which R_{max} and R_{min} are the distances separating the bottom and the top of the liquid column from the axis during centrifugation; $g_{av} = \frac{1}{981}\ \omega^2\ R_{av}$.

Table 2. Reference Enzyme Content of an L Fraction. Percentage values are given with respect to the homogenate. The relative specific activity corresponds to the ratio of the specific activity found in the fraction to the specific activity found in the homogenate.

Enzymes	%	Relative specific activity
Catalase	24.1	7.5
Urate oxidase	36.9	11.5
Acid phosphatase	35.1	10.9
β-Galactosidase	21.5	6.7
Glucose-6-phosphatase	0.7	0.2
NADPH cytochrome c reductase	2.8	0.9
Alkaline phosphodiesterase	5.2	1.6
Galactosyltransferase	0.9	0.3
Cytochrome oxidase	2.9	0.9
Proteins	3.2	–

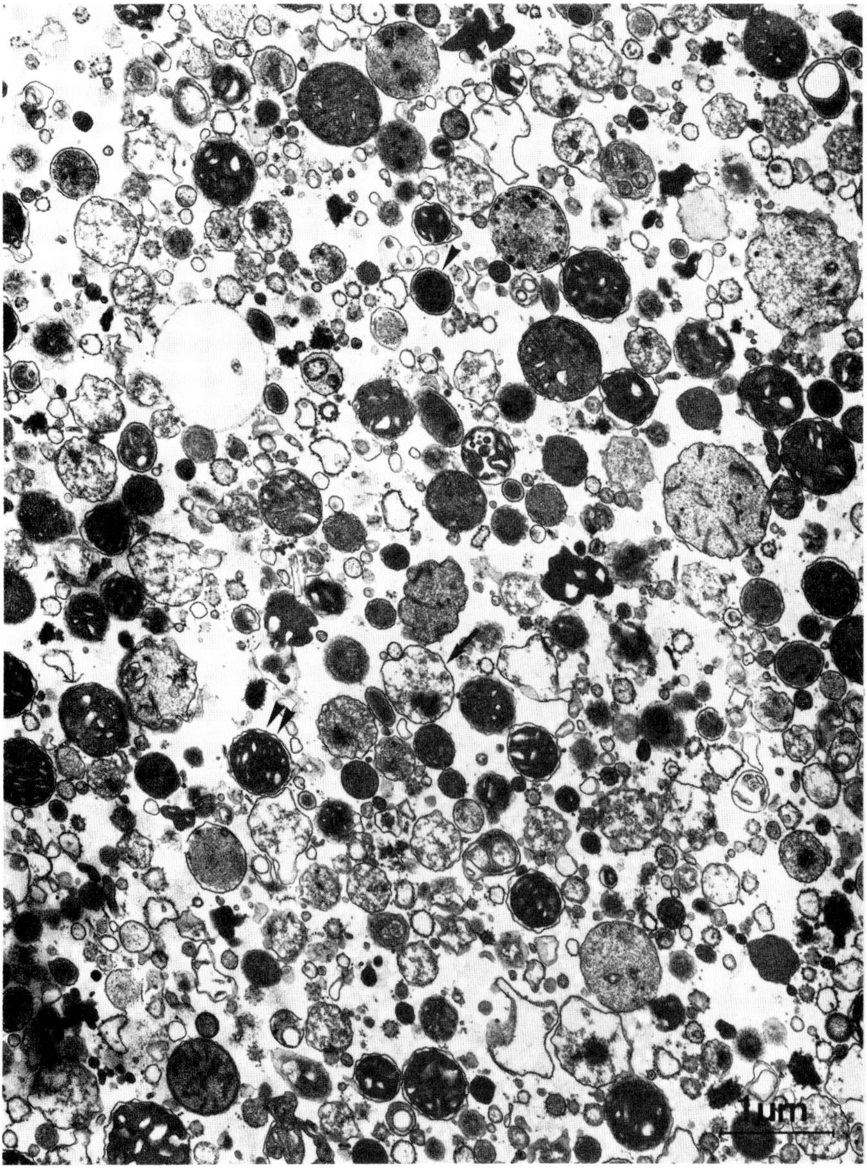

Figure 11. Morphological appearance of an L fraction. (⯈⯈): mitochondria; (►): lysosomes; (→): peroxisomes.

3.2.1. *Preparation of an L Fraction (Light Mitochondrial Fraction)*

The preparation of the light mitochondrial fraction is performed according to de Duve *et al.* (3) as shown in *Figure 10*. The reference enzyme content of such a fraction is exemplified by *Table 2* and its morphological appearance is shown in *Figure 11*.

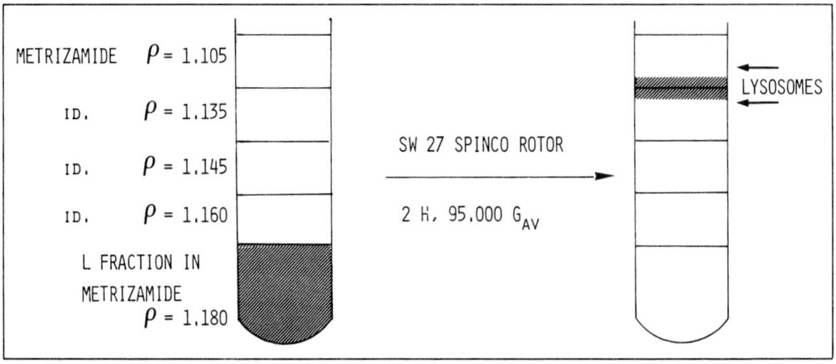

Figure 12. Schematic representation of the discontinuous metrizamide gradient used to purify lysosomes.

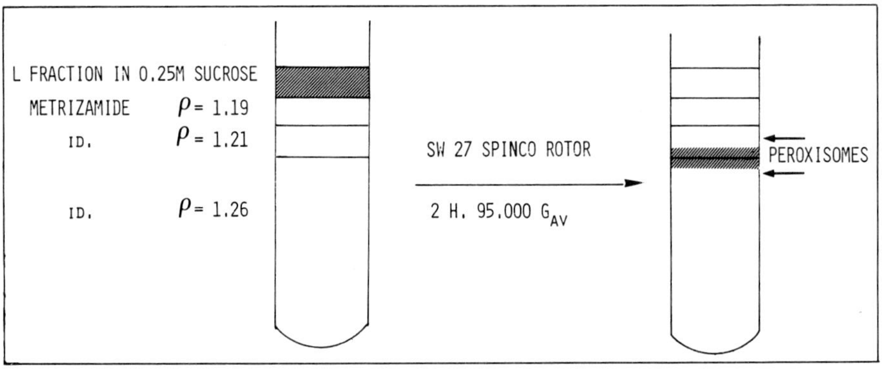

Figure 13. Morphological appearance of a purified lysosome preparation.

Table 3. Reference Enzyme Content of Lysosomal Preparations. The lysosomes were prepared according to Wattiaux *et al.* (6). Percentage values are given with respect to the homogenate. The relative specific activity corresponds to the ratio of the specific activity found in the fraction to the specific activity found in the homogenate.

Enzymes	%	Relative specific activity
Acid phosphatase	12.0	80.0
β-Galactosidase	9.5	64.1
N-acetylglucosaminidase	11.1	66.5
Monoamine oxidase	0.10	0.56
Cytochrome oxidase	0.03	0.17
NADPH cytochrome c reductase	0.04	0.27
Glucose-6-phosphatase	0.04	0.23
Alkaline phosphodiesterase	1.27	8.8
Alkaline phosphatase	1.42	8.7
5′-nucleotidase	0.88	5.4
Catalase	0.02	0.13
Galactosyltransferase	0.3	–
Proteins	0.15	–

3.2.2. *Preparation of Lysosomes*

The method described is similar to the one already published (6) with minor modifications. Prepare the L fraction *(Figure 10)* and dilute it with 0.25 M sucrose to a volume of 1 ml for each 3 g of liver. Add 1 volume of this fraction to 2 volumes of 45% (w/v) metrizamide solution (density 1.25 g/ml). Layer 11 ml of this preparation into the bottom of a 38 ml centrifuge tube (SW 27 tube or equivalent). Then, successively overlayer 6 ml of 30% (w/v) metrizamide (density 1.16 g/ml), 6 ml of 26% (w/v) metrizamide (density 1.145 g/ml), 7 ml of 24% (w/v) metrizamide (density 1.135 g/ml) and 7 ml of 19% (w/v) metrizamide (density 1.105 g/ml). Adjust the density of the solutions by refractometry, using the relationship:

$$\text{Density (g/ml)} = 3.453\eta - 3.601 \qquad (7)$$

Centrifuge the gradients for 2 h at 95 000xg (r_{av} of 11.8 cm) in the Spinco SW 27 rotor or equivalent at 4°C. After centrifugation, collect the fractions as shown in *Figure 12* using a tube slicer similar to the one described by de Duve *et al.* (8).

3.2.3. *Preparation of Peroxisomes*

Purified preparations of peroxisomes can be obtained by centrifugation in a discontinuous density gradient according to the following procedure. Prepare an L fraction *(Figure 10)* and dilute it with 0.25 M sucrose up to a volume of 1 ml for each gram of liver. Layer 4 ml of this preparation on the top of a discontinuous gradient made of 4 ml of 34% (w/v) metrizamide (density 1.19 g/ml), 4 ml of 40% (w/v) metrizamide (density 1.21 g/ml) and 24 ml of 47% (w/v) metrizamide (density 1.26 g/ml) in a Beckman SW 27 tube or equivalent. Centrifugation is performed for 2 h at 95 000xg (r_{av} of 11.8 cm) in a Beckman SW 27 rotor or equivalent rotor at 4°C. After centrifugation, collect the fractions as shown in *Figure 13*.

It is to be noted that as a result of the high sensitivity of peroxisomes to hydrostatic pressure (9,10), it is necessary to minimise the radial distance between the top of the liquid column and the region of the gradient where the particles are recovered. Therefore, it is probable that a Beaufay rotor (11) or a vertical rotor (see Appendix to this chapter) should prove very suitable for isolating peroxisomes in metrizamide gradients.

3.2.4. *Biochemical Composition of the Lysosome Preparation*

As determined by the reference enzyme activities *(Table 3)*, lysosome preparations contain 10–12% of the lysosomes of the tissue with a purification of 60–80-fold with respect to the homogenate. The most significant contaminant appears to be the plasma membrane. However, as discussed elsewhere (6), it is not certain that the activity of the enzymes taken as the markers for the plasma membrane is exclusively associated with this membrane fraction. For example, the cytochemical test for 5′-nucleotidase indicates that this enzyme is probably also present in the lysosomal membrane (6). It is therefore likely that the apparent contamination of the lysosome preparation by the plasma membrane is overestimated.

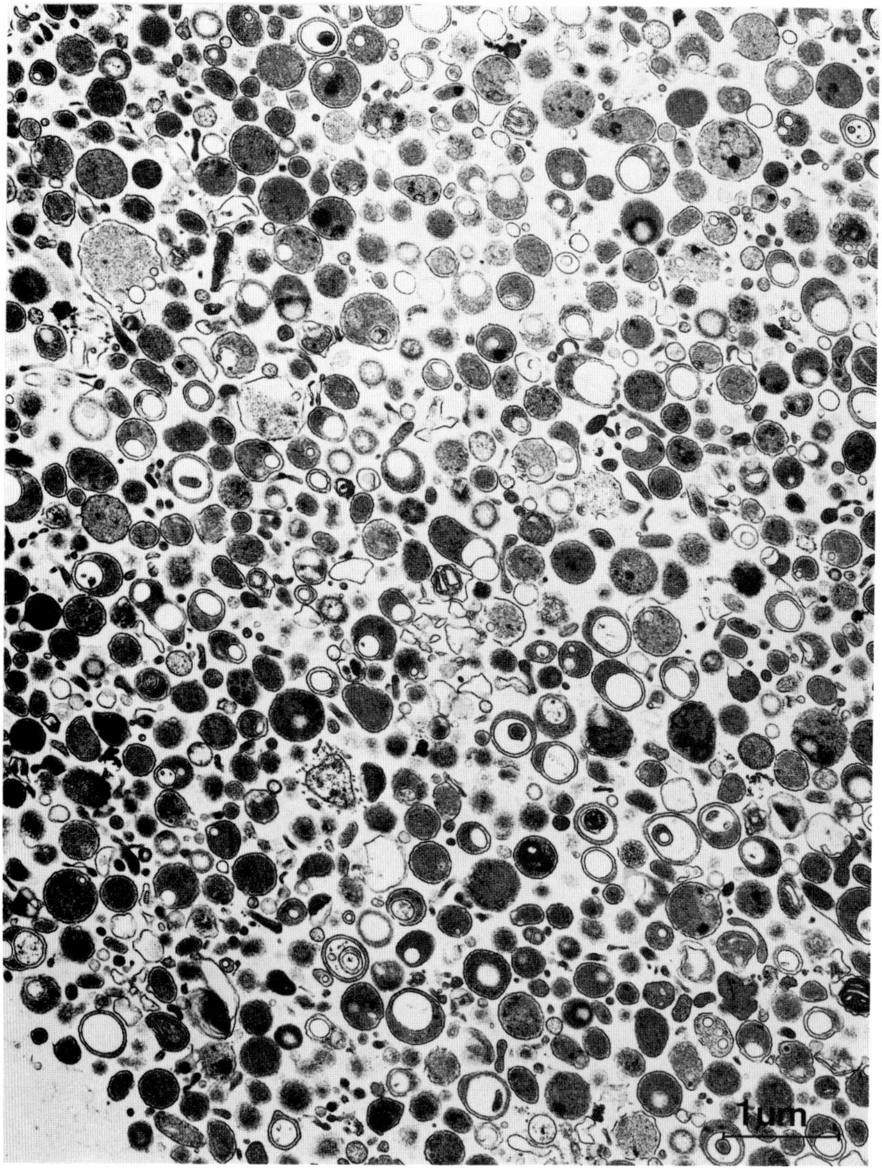

Figure 14. Schematic representation of the discontinuous metrizamide gradient used to purify peroxisomes.

3.2.5. *Morphological Appearance of the Lysosome Preparation*

The general appearance of the purified lysosome preparation is illustrated in *Figure 14*. The main constituents are typical dense bodies with an electronlucent rim beneath their membrane. Most of these particles exhibit a positive reaction to the acid phosphatase cytochemical test (6).

Table 4. Reference Enzyme Content of a Peroxisome Preparation. Percentage values are given with respect to the homogenate. The relative specific activity corresponds to the ratio of the specific activity found in the fraction to the specific activity found in the homogenate.

Enzymes	%	Relative specific activity
Catalase	16.9	30.0
Urate oxidase	19.3	33.2
Acid phosphatase	0.06	0.10
β-Galactosidase	0.09	0.15
Glucose-6-phosphatase	0.18	0.31
NADPH cytochrome c reductase	0.27	0.47
Alkaline phosphodiesterase	0.22	0.38
Galactosyltransferase	0.46	0.79
Cytochrome oxidase	0.049	0.08
Proteins	0.58	–

3.2.6. *Biochemical Composition of the Peroxisome Preparation*

As shown by the activities of catalase and urate oxidase, the peroxisome preparation contains 16 – 20% of the peroxisomes of the tissue with a purification of 30 – 33-fold with respect to the homogenate. Contamination by other membrane components is very low *(Table 4).*

3.2.7. *Morphological Appearance of the Peroxisome Preparation*

As shown in *Figure 15*, the main constituents of the preparation are peroxisomes as described by Leighton *et al.* (13); a few isolated cores, originating from disrupted organelles are also present.

3.3. **Advantages of the Metrizamide Procedure for Isolating Lysosomes and Peroxisomes**

The method most frequently used to isolate lysosomes and peroxisomes is the one described by Trouet (12) and Leighton *et al.* (13). It involves the administration of Triton WR 1339, a non-ionic detergent, to rats. This detergent is picked up by the liver and accumulates in lysosomes (14). As a result, the density of the lysosomes is considerably decreased and the organelles are well separated from the mitochondria and peroxisomes after isopycnic centrifugation in a sucrose gradient (13). As the main contaminants of peroxisomes are lysosomes, the procedure is applicable for purifying peroxisomes. This method, although very useful in many cases, has some drawbacks, particularly when studies on lysosomal membranes have to be performed. Lysosomes containing Triton WR 1339 (tritosomes) do not behave like normal lysosomes in certain circumstances. For example, the free hydrolase activity is higher in a tritosome preparation than in lysosomes isolated in metrizamide gradients. Also, it has been shown that more ATPase activity is associated with tritosomes than with normal lysosomes (15). In the case of peroxisomes there is no real evidence that these granules are affected by the Triton treatment. However, because these organelles are involved in lipid metabolism and since Triton WR 1339 also affects lipid metabolism, it raises the possibility that

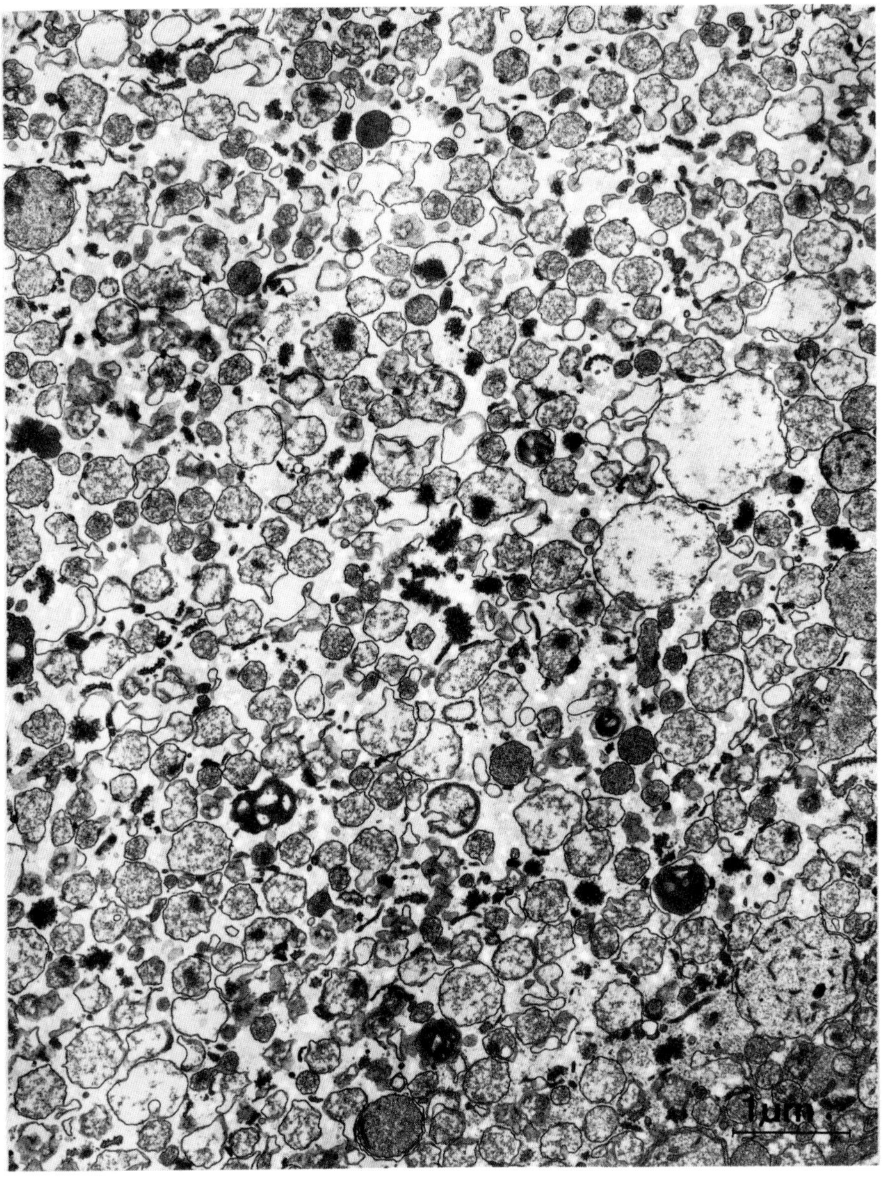

Figure 15. Morphological appearance of a purified peroxisome preparation.

peroxisomes might be affected by the Triton treatment. In any case, the Triton WR 1339 cannot be used when it is not possible to give an injection of detergent. For example, this is the case if lysosomes or peroxisomes have to be isolated from human livers.

It should also be mentioned that Marzella *et al.* (16) recently made use of centrifugation in a discontinuous metrizamide gradient to isolate rat-liver autophagic vacuoles. These structures are directly related to the lysosomal components.

3.4. Isolation of Mitochondria

It does not appear that isopycnic centrifugation in metrizamide gradients would be particularly suitable for isolating mitochondria since, in this case, metrizamide has no advantages over sucrose. As shown in this chapter the overlapping of the distribution curves of marker enzymes of mitochondria and lysosomes is not normally less pronounced in a metrizamide gradient than in a sucrose gradient, unless the mitochondria are subjected to a high hydrostatic pressure during centrifugation. Obviously, such treatment involves alterations of the mitochondrial structure and is certainly not to be recommended unless the method is first rigorously proven.

4. BEHAVIOUR OF PARTICLES IN NYCODENZ GRADIENTS

Nycodenz is an iodinated compound related to metrizamide (see Chapter 1). The physicochemical properties of this substance are similar to those of metrizamide. Therefore, it may be expected that the considerations formulated in this paper for metrizamide can be applied to Nycodenz with relatively minor modifications. A detailed study of the isolation of peroxisomes in self-forming Nycodenz gradients in a vertical rotor has been carried out and is described in the Appendix to this chapter.

5. REFERENCES

1. Beaufay,H. and Berthet,J. (1963) in *Methods of Separation of Subcellular Structural Components* (Grant,J.K. ed.) Biochem. Soc. Sympos. no. 23, Cambridge University Press, Cambridge, p. 65.
2. Beaufay,H., Jacques,P., Baudhuin,P., Sellinger,O.Z., Berthet,J. and de Duve,C. (1964) *Biochem. J.,* **92**, 184.
3. de Duve,C., Pressman,B.C., Gianetto,R., Wattiaux,R. and Appelmans,F. (1955) *Biochem. J.,* **60**, 604.
4. Wattiaux,R., Wattiaux-De Coninck,S. and Ronveaux-Dupal,M.F. (1971) *Eur. J. Biochem.,* **22**, 31.
5. Wattiaux,R. (1974) *Mol. Cell. Biochem.,* **4**, 21.
6. Wattiaux,R., Wattiaux-De Coninck,S., Ronveaux-Dupal,M.F. and Dubois,F. (1978) *J. Cell Biol.,* **78**, 349.
7. Birnie,G.P., Rickwood,D. and Hell,A. (1973) *Biochim. Biophys. Acta,* **331**, 283.
8. de Duve,C., Berthet,J. and Beaufay,H. (1959) *Progr. Biophys. Biophys. Chem.,* **9**, 325.
9. Wattiaux,R., Wattiaux-De Coninck,S., Collot,M. and Ronveaux-Dupal,M.F. (1972) *Spectra,* **4**, 64.
10. Bronfman,M. and Beaufay,H. (1973) *FEBS Lett.* **36**, 163.
11. Beaufay,H., Amar-Costesec,D., Thines-Sempoux,M., Wibo,M. Robbi,M. and Berthet,J. (1964) *J. Cell. Biol.,* **61**, 213.
12. Trouet,A. (1964) *Arch. Int. Physiol. Biochim.,* **72**, 698.
13. Leighton,F., Poole,B., Beaufay,H., Baudhuin,P., Coffey,J.W., Fowler,S. and de Duve,C. (1968) *J. Cell Biol.,* **37**, 482.
14. Wattiaux,R., Wibo,M. and Baudhuin,P. (1963) in *Ciba Foundation Symposium on Lysosomes* (de Reuck,A.V.S. and Cameron,M.P. eds) J. and A. Churchill, Ltd, London, p. 176.
15. Burnside,J. and Schneider,D.L. (1982) *Biochem. J.,* **204**, 525.
16. Marzella,L., Ahlberg,J. and Glaumann,H. (1982) *J. Cell Biol.,* **93**, 144.

Isolation of Peroxisomes Using a Vertical Rotor

H. Osmundsen

1. CHARACTERISTICS OF VERTICAL ROTORS

During the last few years vertical rotors have become commercially available. This type of rotor has a number of characteristics which should make them especially useful for work involving the fractionation of subcellular organelles. Perhaps the most useful characteristic is that the total centrifugal force required to achieve adequate separation can be a sixth, or less, of that required with swinging-bucket rotors (see *Table 1*). Since the sedimentation pathlength is relatively short, in comparison with the gradient volume, a smaller hydrostatic pressure will be generated in the density gradient as compared with swing-out rotors. This should minimise the possibility of damage to organelles as a result of the hydrostatic pressure, which can readily occur in a swing-out rotor (1). Peroxisomes are particularly sensitive to physical stress and may readily rupture, releasing soluble enzymes. The combined effects of the much smaller hydrostatic pressure gradient, and the shorter time of centrifugation, should enable the isolation of peroxisomes which have been subjected to significantly less physical stress during the isolation. The characteristics of vertical rotors have been extensively described by Rickwood (2).

2. METHODS FOR THE ISOLATION OF PEROXISOMES

2.1. Formation of Gradients

With vertical rotors it is possible to self-form density gradients prepared from metrizamide or Nycodenz solutions after relatively short centrifugation runs. Gradients formed by centrifuging homogeneous solutions of Nycodenz for various times up to 120 min are shown in *Figure 1*. These results show that potentially useful gradients are formed even after centrifugation for 60 min at 200 000xg. In this experiment a rotor which can take eight gradients, each of 33 ml volume, was used giving a total gradient volume of just over 250 ml. Using a smaller gradient volume with a shorter sedimentation pathlength and faster maximum speed (e.g. TV 865 or VTi65) a useful gradient should be obtained after about 30 min of centrifugation.

The use of self-forming gradients is primarily time saving, because a large number of gradients can be prepared simultaneously, during one centrifugation run. Also, many different gradients can be prepared simply by varying the starting concentration of Nycodenz. The gradient profiles may of course also

Table 1. Centrifugation Conditions Used to Isolate Peroxisomes.

Gradient medium	*Type of rotor*				*References*
	Swing-out		*Vertical*		
	RCF	*Time (h)*	*RCF*	*Time (h)*	
Sucrose	59 000	4.0	52 000	0.5	8,6
Ficoll[a]	128 000	2.0	n.a.		9
Metrizamide	108 000	2.5	130 000	1.0	1,3
Nycodenz	n.a.		63 000	0.5	10

[a]For Ficoll no procedure which specifically refers to isolation of peroxisomes appears to be available.

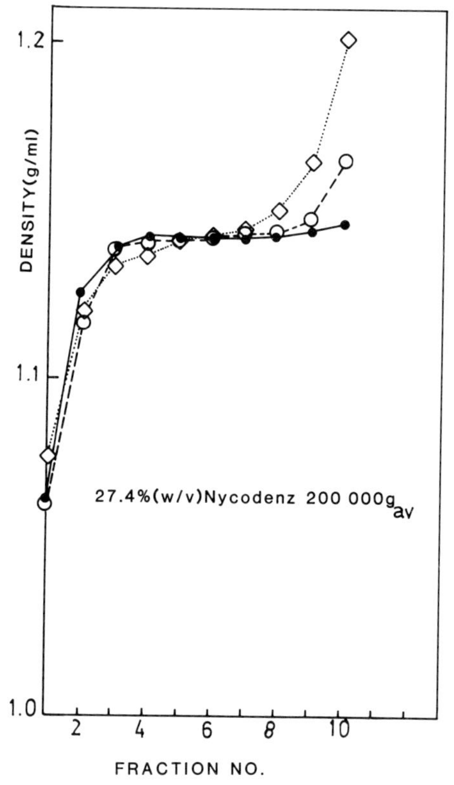

Figure 1. Density profiles generated by centrifugation of a 27.4% (w/v) solution of Nycodenz for 30 min (●), 60 min (○) and 120 min (◇) at 200 000xg using a Sorvall TV 850 rotor. The composition of the medium is described in the text and the volume was 31 ml of 27.4% (w/v) Nycodenz plus 2 ml of 60% (w/v) sucrose to function as a bottom cushion. After fractionation, the density of the various fractions was measured by using a refractometer.

be altered by changing the time of centrifugation, or the speed of centrifugation (see Chapter 1 section 4.2). In practice, however, the TV 850 (or VTi50) rotor is usually run at its maximum speed to keep the time required for gradient generation to a minimum (see *Figure 1*).

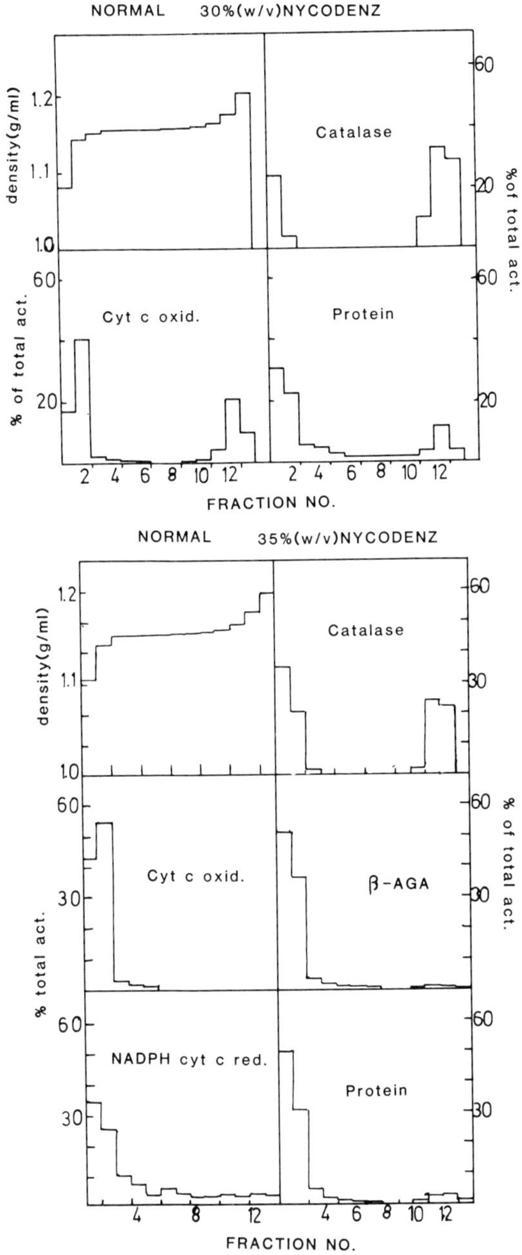

Figure 2. Distribution of marker enzyme activities after centrifugation of L-fractions isolated from normal rat livers on gradients formed from (a) 30% (w/v) or (b) 35% (w/v) of Nycodenz. Centrifugation was carried out for 30 min at 63 000xg using a Sorvall TV 850 rotor. After fractionation of the gradients, density and enzyme activities were measured in the various fractions.

The density gradients used in the procedures described here contained 30% or 35% (w/v) Nycodenz, 1 mM EGTA, 10 mM Mops-KOH (pH 7.2), and 0.1% ethanol. For pre-generation of the gradient pour 31 ml of this medium

Table 2. Preparation of an L-fraction From Rat Liver.

1. Homogenise the liver of a freshly killed rat, using a Potter-Elvehjem homogeniser, in a medium containing 300 mM mannitol, 1 mM EGTA and 10 mM Hepes-KOH (pH 7.2). Use small portions of finely chopped liver each time, using a single stroke with the homogeniser pestle to achieve adequate homogenisation. Dilute the final homogenate to 10% (w/v) prior to further use.
2. Centrifuge the homogenate at 5900xg for 5 min at 3°C to remove cell debris, nuclei and a large fraction of the mitochondria.
3. Discard the pellet and centrifuge the supernatant at 41 000xg for 10 min. Resuspend the resulting pellet, using a glass homogeniser, in about 3 ml of the homogenisation medium containing, in addition, 0.1% (v/v) ethanol and 5 mM EGTA.
4. Centrifuge the resulting suspension for 3 min at 1050xg to remove material which is not completely disaggregated. The resulting supernatant constitutes the L-fraction. Usually 1 ml is layered onto 30 ml gradients; this is about 50 mg of protein. This fraction is enriched with respect to peroxisomes, but also contains appreciable amounts of lysosomes, endoplasmic reticulum as well as mitochondria. For further details see ref. 4.

into a centrifuge tube, place a cushion of 2 ml of 60% (w/w) of sucrose into the bottom of the tube, and seal the tube. Generate the gradient by centrifugation at 200 000xg for 60 min.

Peroxisomes have also been purified on preformed metrizamide gradients using a vertical rotor (3). To form a linear 20 – 50% (w/v) metrizamide gradient, prepare two solutions of 20% and 50% (w/v) metrizamide and use a simple two-chambered gradient maker. Alternatively, prepare step gradients by overlayering 8 ml each of solutions containing 20%, 30%, 40% and 50% (w/v) metrizamide and allowing them to diffuse to linearity either overnight in a vertical position or for 2 h if the tube is sealed and laid on its side.

2.2. Loading and Fractionation of Gradients

For the fractionations described in this appendix L-fractions were used. *Table 2* gives the standard method used to prepare L-fractions. Although the L-fraction is depleted of mitochondrial material, the mitochondrial contribution to the L-fraction is still very substantial compared with the peroxisomal content.

Prior to loading self-formed gradients, remove the top 2 ml carefully using a graduated pipette, and replace it by 1 ml of sample followed by 1 ml of overlay solution (100 mM mannitol, containing 0.1 mM EGTA, 5 mM Hepes-KOH, pH 7.2). It is usually good practice to include an overlay when using a vertical rotor, to prevent possible interaction between the wall of the centrifuge tube and sample, which may cause loss of resolution. After the gradients are loaded with the sample centrifuge them at 63 000xg for 30 min. After centrifugation, the gradients are immediately fractionated by pumping Maxidens (Nyegaard & Co.) to the bottom of the centrifuge tube and pushing the gradient out of the tube, the top of the gradient emerging first.

2.3. Purification of Peroxisomes on Self-forming Nycodenz Gradients

L-fractions derived from normal rat livers, and from livers of rats treated with clofibrate (0.3%, w/w) in the diet were used (clofibrate is known to cause proliferation of hepatic peroxisomes, ref. 5). *Figure 2* shows the typical marker

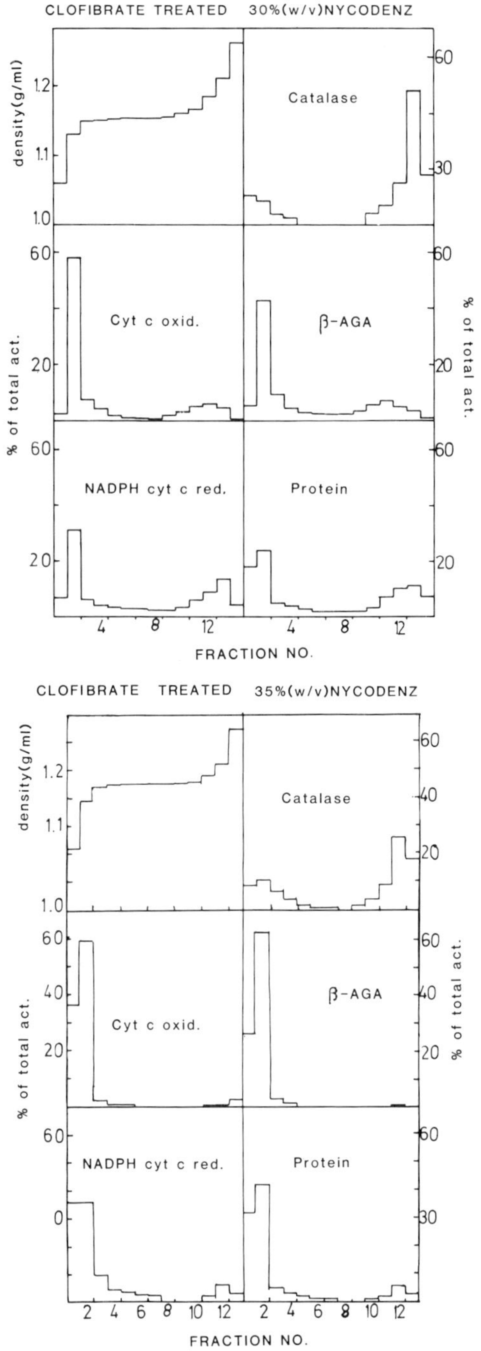

Figure 3. Distribution of marker enzyme activities after centrifugation of L-fractions prepared from liver of rats which had been treated with clofibrate for 12 days. The gradients were formed from (a) 30% or (b) 35% (w/v) Nycodenz solution. Centrifugation was carried out for 30 min at 63 000xg using a Sorvall TV 850 rotor.

Table 3. Enzymatic Activities Associated with Purified Peroxisomal Fractions.

Enzyme activity	*Type of gradient*			
	30% (w/v) Nycodenz		*35% (w/v) Nycodenz*	
	Control	*Clofibrate*	*Control*	*Clofibrate*
Catalase	72%	81% (69)	48% (75)	65% (84)
Cytochrome c oxidase	16%	22%	ND	1.6%
Rotenone insensitive NADPH-cytochrome c reductase	31%	35%	11%	9%
β-N-acetyl-D-glucosaminidase	22%	27%	4%	3%
Protein (Bio-Rad assay)	24%	40%	9%	14%

The amounts of the various marker enzyme activities found in fractions 12 and 13 of *Figures 2* and *3* have been expressed as a percentage of the total amount of activity found in the fractions. The numbers in parentheses represent estimates of purity of these fractions, expressed as percentages of peroxisomal protein. These fractions were obtained by layering L-fractions from normal rats or from rats treated with clofibrate for 12 days onto gradients self-generated from 30% (w/v) or 35% (w/v) solutions of Nycodenz, as described in the text. The purity of the peroxisomal fractions was calculated using the values of Leighton *et al.* (4).

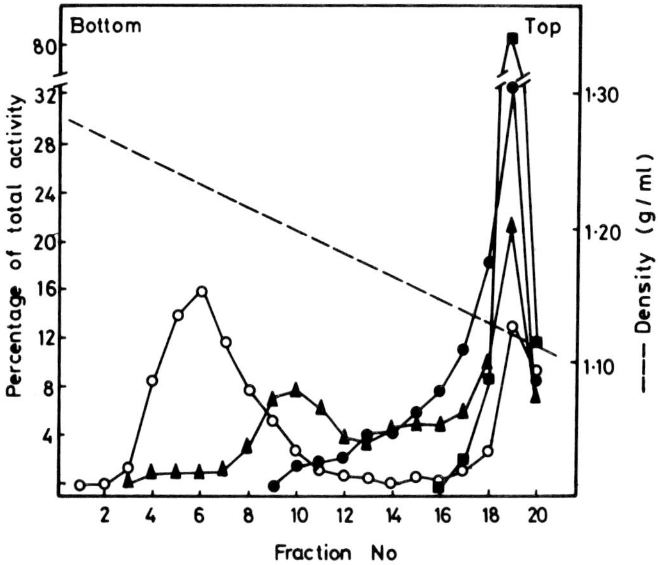

Figure 4. Distribution of marker enzyme activities after centrifugation of a liver L-fraction on a 20 – 50% (w/v) metrizamide gradient. An L-fraction derived from guinea pig liver was loaded onto a preformed 20 – 50% (w/v) metrizamide gradient. The gradient was centrifuged at 130 000xg for 60 min in a Beckman VTi 50 rotor. The distributions of catalase (○——○), glucose-6-phosphatase (▲——▲), acid phosphatase (●——●) and succinate-cytochrome c reductase (■——■) were measured. The density profile (-----) was determined by refractometry. Data derived from ref. 3.

enzyme profiles obtained with L-fractions from normal rats, which had been centrifuged in 30% and 35% (w/v) Nycodenz self-forming gradients. With gradients derived from 30% (w/v) Nycodenz solutions most of the catalase activity in the gradient sediments to a buoyant density of 1.22 g/ml, which is close to the density of peroxisomes in metrizamide gradients (3). Appreciable amounts of mitochondria, lysosomes and endoplasmic reticulum are also present in these peroxisomal fractions, as judged by the distributions of cytochrome c oxidase, β-N-acetyl-D-glucosaminidase, and rotenone-insensitive NADPH cytochrome c reductase, respectively. This applies to samples obtained from liver homogenates from control animals as well as from livers of rats treated with clofibrate. However, when using self-forming gradients derived from 35% (w/v) Nycodenz solutions, the extent of contamination of the peroxisomal fractions with the enzymic activities associated with these other organelles is markedly decreased (see *Figure 3*). With L-fractions from control rats cytochrome c oxidase activity was not detectable in the peroxisomal fractions, although a very low level of activity (about 2% of the total activity in the gradient) was detectable in the peroxisomal fractions obtained from L-fractions of rats treated with clofibrate. With fractions from control rat livers the content of catalase in fraction 1 was always higher than with L-fractions from livers of rats treated with clofibrate, irrespective of whether 30% or 35% Nycodenz gradients had been used. This suggests that peroxisomes from normal rat livers are more fragile than those isolated from treated rats.

As can be seen from the data presented in *Table 3*, peroxisomal fractions isolated in these gradients can contain up to about 85% of peroxisomal protein, the major contaminant being endoplasmic reticulum. However, on dilution of the peroxisomal fraction followed by centrifugation most of the contaminating endoplasmic reticulum remains in the supernatant separate from the peroxisomes which are pelleted. A similar phenomenon has previously been shown to occur with similar fractions isolated in sucrose or Percoll density gradients (6,7). The use of self-forming Nycodenz gradients therefore constitutes a very rapid, and convenient, procedure for isolation of rat-liver peroxisomes.

2.4. Purification of Peroxisomes on Preformed Metrizamide Gradients

Fractionation of an L-fraction from liver on a preformed 20 – 50% (w/v) metrizamide gradient also gives very pure peroxisomal material *(Figure 4)*. As judged by the marker enzymes for lysosomes (acid phosphatase) and mitochondria (succinate-cytochrome c reductase) neither of these organelles contaminate the peroxisomal fractions. In addition, in these metrizamide gradients the bulk of the glucose-6-phosphatase activity which is associated with the endoplasmic reticulum bands at a lower density than the peroxisomes *(Figure 4)*.

3. CONCLUSIONS

The data given in this appendix emphasise the excellent degree of purification of peroxisomes that can be achieved using either self-forming or preformed gradients of both Nycodenz and metrizamide. Moreover, using a vertical rotor large quantities of pure peroxisomes can be prepared in a very short time. As compared with sucrose gradients, the methods described here are simpler, expose the organelles to less osmotic stress and give purer preparations than when sucrose gradients are used. In addition, unlike sucrose gradients, one can obtain very high yields of peroxisomes using these methods. One can prepare peroxisomes which, in contrast to fractions obtained in self-formed Percoll gradients, are essentially free of contaminating mitochondria. Hence, metrizamide and Nycodenz gradients should prove most useful in studies of the localisation of peroxisomal enzymes.

4. REFERENCES

1. Wattiaux,R., Wattiaux-De Coninck,S., Roneveaux-Dupal,M.F. and Dubois,F. (1978) *J. Cell Biol.,* **78**, 349.
2. Rickwood,D. (1982) *Anal. Biochem.,* **122**, 33.
3. Hajra,A.K. and Bishop,J.E. (1982) *Ann. N.Y. Acad. Sci.,* **386**, 170.
4. Leighton,F., Poole,B., Beaufay,H., Baudhuin,P., Coffey,J.W., Fowler,S. and de Duve,C. (1968) *J. Cell Biol.,* **37**, 482.
5. Svoboda,J., Grady,H. and Azarnoff,D. (1967) *J. Cell Biol.,* **35**, 127.
6. Neat,C.E. and Osmundsen,H. (1979) *Biochem. J.,* **180**, 445.
7. Neat,C.E., Thomassen,M.S. and Osmundsen,H. (1981) *Biochem. J.,* **196**, 149.
8. Wattiaux-De Coninck,S. and Wattiaux,R. (1971) *Eur. J. Biochem.,* **19**, 552.
9. Beaufay,H., Jacques,P., Baudhuin,P., Sellinger,O.Z., Berthet,J. and de Duve,C. (1964) *Biochem. J.,* **92**, 184.
10. Osmundsen,H. and Cervenka,J. (1982) *Hoppe-Seyler's Z. Physiol. Chem.,* **363**, 1000.

CHAPTER 7

Fractionation of Mammalian Cells

A.BØYUM, T. BERG and R. BLOMHOFF

1. GENERAL INTRODUCTION

The sedimentation rate of cells in a liquid depends on the size and density of the cells as well as the density and viscosity of the surrounding medium. However, in the case of mammalian cells the osmolarity of the medium also affects the sedimentation rate and density of the cells in that, as the osmolarity of the surrounding medium becomes hypertonic, the cells lose water and shrink. Common gradient media such as sucrose and CsCl have high osmotic potentials (see Chapter 1 Section 3) and hence solutions of these compounds, at the densities required for the isopycnic banding of cells, are extremely hypertonic. Solutions which are very hypertonic will not only affect the size and density of the cells but also may reduce their viability. One approach to this problem is to use macromolecular media (e.g. Ficoll or bovine serum albumin) or particulate media (e.g. Ludox or Percoll) that have low osmotic potentials. The uses of these media have been reviewed elsewhere (1). Macromolecular gradient media have high viscosities while those based on colloidal silica have been shown to interact with some types of cells. As can be seen from their physico-chemical properties (see Chapter 1 Section 3), nonionic iodinated density-gradient media are suitable for separating cells. Although iodinated density-gradient media do have an osmotic potential, it is possible to form isotonic gradients of sufficient density using either NaCl or sucrose as an osmotic balancer (Chapter 1 Section 4.1). The other important features of the nonionic iodinated gradient media are that extensive testing has shown that these media do not affect the morphology of cells and neither are they toxic to cells (2).

This chapter describes the use of gradients of iodinated media for the separation of different types of disaggregated liver cells and of blood cells as examples of the types of technique that can be used to fractionate a wide variety of animal cells. As described in this chapter, isotonic gradients of Nycodenz and metrizamide can not only be used to fractionate various types of cells but also it is possible to separate viable and non-viable cells. In addition, it is found that, on occasions, the fractionation of cells can be optimised by using mildly hypertonic conditions.

2. ISOLATION AND FRACTIONATION OF LIVER CELLS

2.1. **Introduction**

Isolated liver cells can be prepared by treating the perfused liver with collagenase (see Appendix to this chapter). The cells obtained are viable and in very high yield and they are an ideal starting material for the separation of the various types of liver cells (3). Rat liver contains in addition to its parenchymal cells (also called "hepatocytes") at least three main types of non-parenchymal cells: Kupffer cells (macrophages), endothelial cells and stellate cells (fat storing cells). The parenchymal cells make up about 90% of the liver mass and about 65% of the total cell number (4). The non-parenchymal cells are mainly the sinusoid lining cells. These cells are much smaller than the parenchymal cells and consist of about 60% endothelial cells, 30% Kupffer cells and 10% stellate cells (5).

2.2. Separation of Intact and Damaged Liver Cells

During tissue disaggregation or subsequent fractionation of the cells, some cells become damaged and non-viable usually as a result of damage to the cell membrane. Seglen and coworkers first reported (15) that it was possible to separate intact and damaged, non-viable cells using a discontinuous metrizamide gradient; a diagrammatic representation of this technique is shown in *Figure 1*. The basis of this separation is that damaged cells are freely permeable to the surrounding medium and hence the intracellular water does not contribute to the overall density of the cell. The result of this is that damaged cells band denser than viable cells in metrizamide gradients *(Figure 1)*; in part a reflection of the presence of the dense nucleus of the cell.

The use of Nycodenz for the separation of viable and non-viable non-parenchymal liver cells has also been investigated. In Nycodenz gradients liver cells band over the range 1.06 – 1.12 g/ml and it has been found that it is possible to separate viable and non-viable cells as follows. Mix three volumes of the cell suspension (10^7 cells/ml) in isotonic medium containing 0.15 M NaCl, 5 mM KCl, 0.4 mM Na_2HPO_4, 0.3 mM KH_2PO_4, 1.0 mM $MgSO_4$, 2.0 mM $CaCl_2$ and 20 mM Hepes-NaOH (pH 7.5) (see Appendix to this chapter) with four volumes of 40% (w/v) Nycodenz containing 6.5 mM KCl, 1.0 mM $CaCl_2$ and 10 mM Hepes-NaOH (pH 7.5) to give a final concentration of 22% (w/v) Nycodenz. Carefully overlayer the cell suspension with 5 ml of the isotonic medium and centrifuge the discontinuous gradient at 500xg for 5 min at 15°C. The damaged, non-viable cells are pelleted while the viable cells float up to the interface and can be collected using a Pasteur pipette.

2.3. Separation of Parenchymal and Non-parenchymal Cells

2.3.1. *Use of Differential Centrifugation*

So far no one technique is capable of separating the various types of liver cell in a single step, for instance by isopycnic centrifugation. Instead various properties of the cells have been utilised to separate the cells using several methods. The main problem in separating liver cells is to obtain the non-parenchymal cells in high yield and to subfractionate these cells into the three main groups of cells. The parenchymal cells are easily separated from the non-parenchymal cells by differential centrifugation (6). The average volume of the parenchymal cell is more than ten times that of the average non-parenchymal

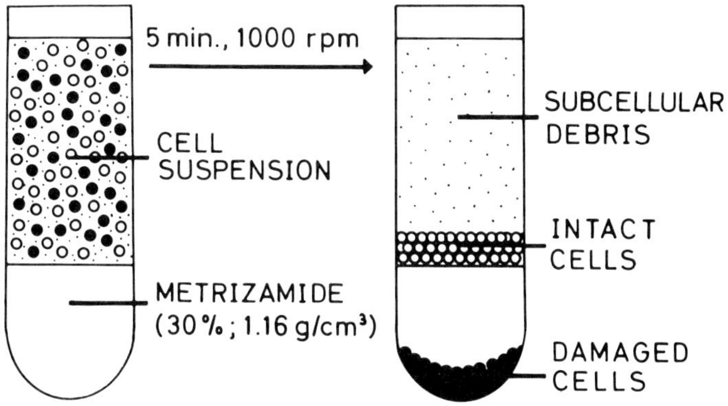

Figure 1. One-step separation of intact from damaged cells. A cell suspension containing a mixture of intact and damaged cells is layered above a cushion of buffered 30% (w/v) of metrizamide (density = 1.16 g/cm^3). The sharp interface is eliminated with a mixing device (a thin metal thread coiled at one end into a flat spiral, bent at an angle of 90° so that the length of the thread serves as a shaft) before centrifugation in order to minimise tight packing of cells. After 5 min of centrifugation at 200xg (1000 rev/min) (a higher centrifugation force may be used if required), the damaged, non-viable cells have sedimented to the bottom of the tube, while the intact cells band at the interface. By removing the supernatant and centrifuging a second time at higher speed, the subcellular debris in the intact cell layer also moves into the metrizamide cushion, and a very clean preparation of intact cells can be obtained. From ref. 15.

cell, and centrifugation of a liver cell suspension at low speed (about 50xg) for 15 sec at 4°C gives a fairly good separation of the parenchymal cells and the non-parenchymal cells. The pellet, containing the bulk of the parenchymal cells, can be washed repeatedly to give a pure suspension of hepatocytes. The supernatant is, of course, enriched in non-parenchymal cells but still contaminated with some hepatocytes. Dead hepatocytes do not sediment during low-speed centrifugation and will therefore be selectively retained in the supernatant. Nevertheless, by repeated centrifugation at 600xg for 4 min at 4°C a fairly pure preparation of non-parenchymal cells may be obtained from the initial supernatant after sedimentation of the hepatocytes. The cells are viable as judged by their endocytic activity *in vitro* and by the trypan blue exclusion test (6). However, the yield of cells is relatively low, and Kupffer cells seem to be lost selectively. About 25% of the non-parenchymal cells in the starting material are left in the final preparation. The fraction of Kupffer cells as a percentage of the non-parenchymal cells decreases from 30% to about 15%. The selective loss of Kupffer cells probably occurs because these cells tend to aggregate and sediment with the parenchymal cells.

2.3.2. *Preparation of Non-parenchymal Liver Cells Using Pronase Digestion*

The yield of non-parenchymal cells can be increased by taking advantage of agents which selectively destroy the parenchymal cells. Pronase has long been used for this purpose (7). After incubating a suspension of liver cells for 30–60 min with 0.1–0.25% (w/v) pronase the parenchymal cells are preferentially destroyed and the non-parenchymal cells can be washed free of cell debris. The yield of cells is high; at least 50% of the non-parenchymal cells in the starting material can be recovered (8). The proportions of the various non-parenchymal liver cells are representative of those found in the intact liver suggesting that no selective losses occur during this procedure. The viability of the cells is also high (as judged by the trypan blue exclusion test). However, pronase obviously modifies the cell surface and receptors of various kinds may be lost (9). Pronase prepared non-parenchymal cells attach to tissue culture dishes after being incubated overnight (10). During incubation the cells recover receptors; both C_3 and Fc receptors are present in the cultured cells (10). When non-parenchymal cells are prepared by incubating liver cell suspensions with pronase at 37°C, only Kupffer cells seem to attach and the cultivation step therefore may be considered as a separation or purification step. Knook and coworkers (11) prepare non-parenchymal liver cells after treating the perfused liver with pronase. Using this method all types of non-parenchymal cells attach to the tissue culture dish during cultivation (12).

2.3.3. *Preparation of Non-parenchymal Liver Cells Using Enterotoxin*

Another way of destroying hepatocytes selectively is to treat the liver cell suspension with enterotoxin from *Clostridium perfringens* (13). Enterotoxin at concentrations above 1 μg/ml causes gross damage to the hepatocyte membranes with complete loss of lactate dehydrogenase activity from these cells (14). The non-parenchymal cells are not affected by this treatment. These observations are in agreement with the finding that non-parenchymal cells bind only negligible amounts of the toxin while hepatocytes contain about 10^6 binding sites per cell with an association constant (K_a) about 10^6 M^{-1} (14). Enterotoxin treatment has a number of advantages over the pronase digestion method. Enterotoxin acts more selectively on the hepatocytes and can be used in low concentrations (about 20 nM as compared to 20 μM for pronase), also it has no enzymatic activity (14) and does not modify the surface of the non-parenchymal cells. Freshly prepared cells show normal endocytic activity as measured in terms of the binding and uptake of formaldehyde-treated albumin (13). The method involves incubating a suspension of liver cells, prepared as described in the Appendix, at a concentration of 10^7 cells/ml at 37°C in the presence of 10 μg/ml of enterotoxin from *Clostridium perfringens* for 20–30 min. This treatment makes the parenchymal cells leaky as shown by trypan blue staining but the cells do not disintegrate. It is possible to separate the damaged, non-viable hepatocytes from the viable non-parenchymal cells in medium containing Nycodenz or metrizamide as described in Section 2.2. The non-parenchymal cells can be collected from the top of the dense cushion after centrifugation (16). The yield of non-parenchymal cells from the enterotoxin treated cell suspension is usually about 35 x 10^6 per gram of liver. This recovery of cells is comparable to that obtained using pronase (8).

2.4. **Fractionation of Non-parenchymal Rat-liver Cells**

2.4.1. *Introduction*

So far three methods for the preparation of non-parenchymal rat-liver cells have been described (Sections 2.3.1 – 2.3.3). All of these preparations of non-parenchymal liver cells contain endothelial cells, Kupffer cells and stellate cells. The size and density of these cells are so similar that it is difficult to achieve a good separation of these cells in density gradients. Alternative approaches have been tested. As already mentioned, Kupffer cells can under certain circumstances be purified simply by seeding the non-parenchymal cells prepared by pronase treatment; in this case only the Kupffer cells attach to the culture dish (10). Kupffer cells can also be removed very effectively from mixed cell suspensions by means of a magnet after these cells have been loaded with colloidal iron *in vivo*. In this way a cell suspension devoid of Kupffer cells can be obtained (17).

Centrifugal elutriation has also been used for the fractionation of non-parenchymal cells (16,18). However, when non-parenchymal cells are prepared using pronase digestion the stellate cells co-elute with the endothelial cells while when the enterotoxin method is used the stellate cells co-elute with the Kupffer cells. Pure endothelial cells can be prepared from enterotoxin prepared non-parenchymal cells by elutriation.

2.4.2. *Separation of Non-parenchymal Liver Cells in Nycodenz Gradients*

Attempts have been made to separate the various types of rat liver cells by means of density gradient centrifugation in Nycodenz gradients.

(i) *Preparation and analysis of gradients.* The Nycodenz gradients can be prepared by the gradient formation method of Stone (19) as follows. Prepare four solutions containing 27.6%, 18.3%, 13.8%, and 5% (w/v) Nycodenz by mixing the cell suspension with varying amounts of isotonic Nycodenz solution containing 27.6% (w/v) Nycodenz in 5 mM Tris-HCl (pH 7.5), 3 mM KCl and 0.3 mM $CaNa_2$ EDTA. Carefully layer four 2.5 ml portions of these solutions on top of each other, seal the tubes with a bung or Parafilm, turn the tubes to a horizontal position and leave them at 4°C for 90 min. After 90 min turn the tubes upright and centrifuge the diffusion generated gradients at 15°C for 60 min at 1300xg. After centrifugation fractionate the gradients by upward displacement with Maxidens (Nyegaard & Co., Oslo). The density of each fraction can be determined directly by pycnometry (see Section 3.2.1) or calculated from the refractive index of each fraction. It should be noted that refractive index measurements will tend to give less accurate results.

(ii) *Identification of cells.* Hepatocytes in gradient fractions can be identified by their large size and Kupffer cells by their endogenous peroxidase activity (18). Stellate cells can be identified by their content of ^{3}H-retinol given to the animals 7 – 9 h before sacrifice. The newly-absorbed retinol is taken up specifically by the liver parenchymal cells as part of the chylomicron remnants (16,20). Subsequently the vitamin is transferred by an unknown mechanism from the hepatocytes to the non-parenchymal cells (20). The vitamin released from the hepatocytes is sequestered by the stellate cells and hence ^{3}H-retinol administered in this way should be a reliable marker for stellate cells. Endothelial cells represent the bulk of the peroxidase-negative cells (11), and the distribution of peroxidase-negative cells therefore roughly reflects the distribu-

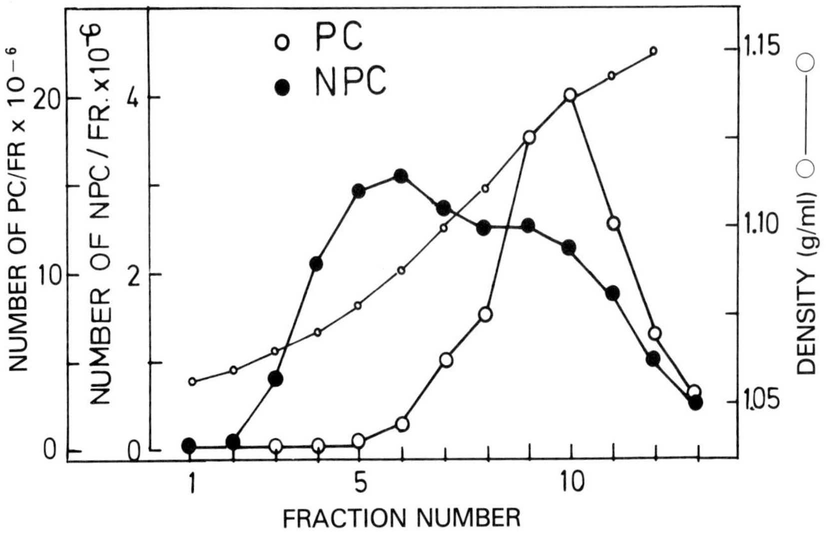

Figure 2. Separation of parenchymal and non-parenchymal rat-liver cells in Nycodenz gradients. The Nycodenz gradients were prepared as described in the text. After centrifugation at 15°C for 60 min at 1300xg (in a Hettich Universal/K2S centrifuge) the gradients were divided into 13 fractions, and the number of parenchymal cells (PC) and non-parenchymal cells (NPC) were counted in each fraction. The density of each fraction was calculated from the refractive index.

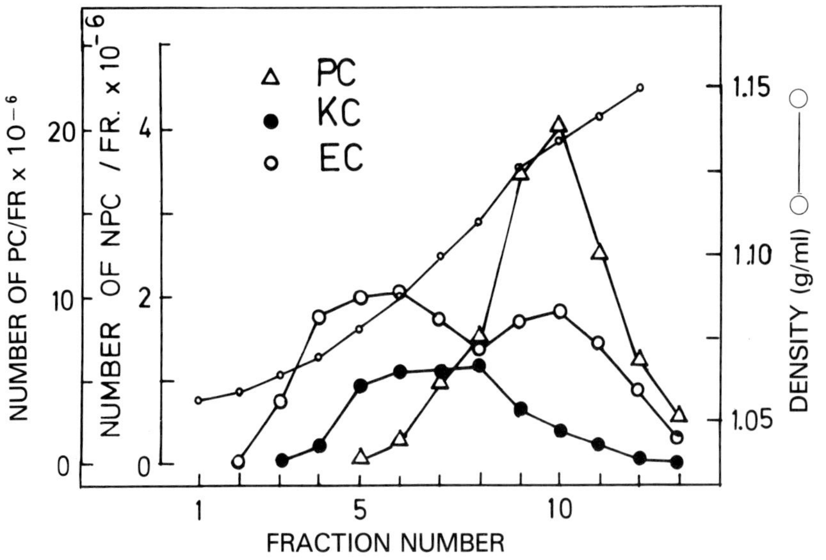

Figure 3. Identification of rat-liver cells separated in Nycodenz gradients. The experiment was performed as described in the legend to *Figure 2*. The 13 gradient fractions (Fr.) were analysed for parenchymal cells (PC), peroxidase-postive Kupffer cells (KC) and endothelial cells (EC).

tion of endothelial cells. Smedsrød and his coworkers have demonstrated that fluorescein-labelled ovalbumin is selectively taken up by the endothelial cells (21). This is probably true also for ^{14}C-sucrose labelled formaldehyde-treated human serum albumin (17). The latter marker has the advantage that its labelled degradation products are trapped in the lysosomes of the target cell.

(iii) *Fractionation of total liver cell suspensions.* The results are depicted in *Figure 2* and show that the median density of the non-parenchymal cells (about 1.09 g/ml) is lower than that of the hepatocytes (about 1.12 g/ml). However, a significant fraction of non-parenchymal cells trail into the same region of the gradient as the hepatocytes; *Figure 3* reveals that these cells are mainly peroxidase-negative cells. Hence, although the separation of hepatocytes and Kupffer cells is fairly good, the degree of contamination of hepatocytes with non-parenchymal cells would appear to be significant. The separation could possibly be improved by diluting the total cell suspension and by introducing changes in the osmolarity of the gradient (see Section 3.5.3).

(iv) *Fractionation of non-parenchymal cells.* Attempts at fractionating non-parenchymal cells prepared by pronase digestion or the enterotoxin method on continuous isotonic Nycodenz gradients have not so far proved to be very successful. Non-parenchymal cells prepared by differential centrifugation would

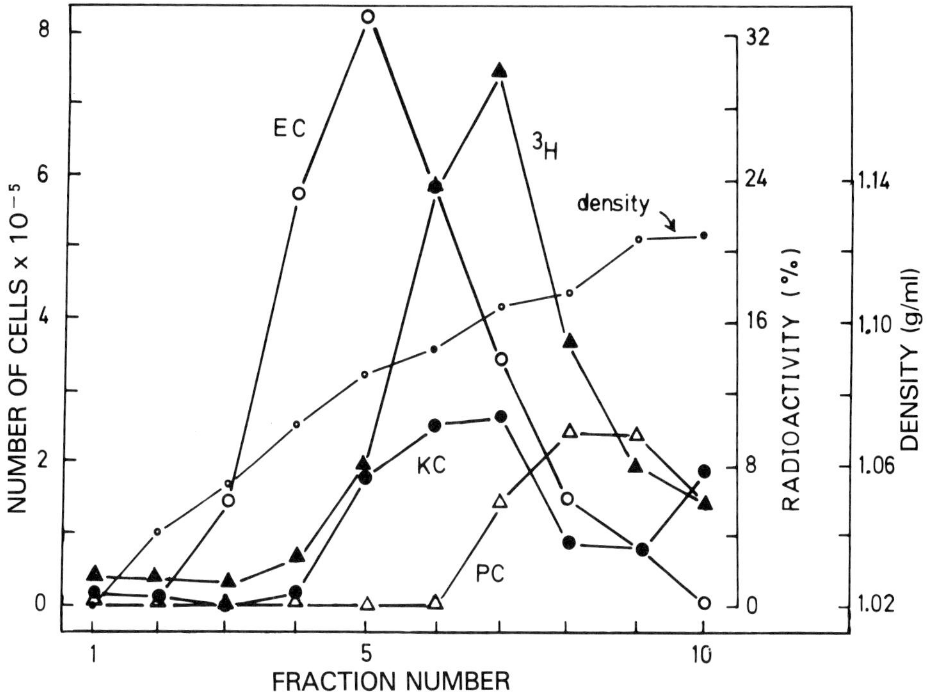

Figure 4. Separation of non-parenchymal rat-liver cells in Nycodenz gradients. Purified non-parenchymal liver cells were prepared by differential centrifugation (6), and introduced into the gradient as described in the text. The gradients were centrifuged at 15°C for 60 min at 1300xg, and then fractionated into 10 fractions. Parenchymal cells (PC) (remaining in the purified fraction of non-parenchymal cells), peroxidase-positive Kupffer cells (KC), and peroxidase-negative cells (mostly endothelial cells) (EC) were counted in the fractions. Radioactivity from ^{3}H-retinol (given to the animals 2–3 h before sacrifice) was also measured in the fractions to determine the distribution of stellate cells.

appear to be a more promising starting material. *Figure 4* shows the distribution of peroxidase-negative cells, Kupffer cells (peroxidase-positive cells) and stellate cells (retinol-containing cells) after centrifugation of non-parenchymal cells in a continuous Nycodenz gradient. A few viable parenchymal cells contaminate the preparation of non-parenchymal cells; their distribution is shown in *Figure 4*. The distributions of cells shown in *Figure 4* suggest that the four types of cells identified band at different densitities as follows: endothelial cells (peroxidase negative) at 1.07 g/ml; Kupffer cells at 1.09 g/ml; stellate cells at 1.09 g/ml and hepatocytes at 1.11 g/ml. The separation of Kupffer cells and endothelial cells is in fact better than indicated in *Figure 4* since a fraction of the denser peroxidase-negative cells are stellate cells. The data shown in *Figure 5* were obtained in an experiment in which the Kupffer cells were removed from the suspension of non-parenchymal cells using a magnet. In this method the Kupffer cells are loaded with colloidal iron particles by injecting each animal with 100 mg of iron carbonyl in 200 μl of 0.9% NaCl into the right femoral vein 30 min prior to perfusion of the liver. A specially constructed magnet (or more correctly: a series of horseshoe magnets, Eclipse, AX510, E.J. Neill and Co., Sheffield, U.K.) is covered with a sheet of Parafilm and about 25 ml of total cell suspension are poured onto the film (8). The cells that do not attach to the magnets are washed twice with 25 ml of isotonic medium. The whole procedure is repeated twice. As can be seen in *Figure 5* the removal of Kupffer cells does not significantly influence the

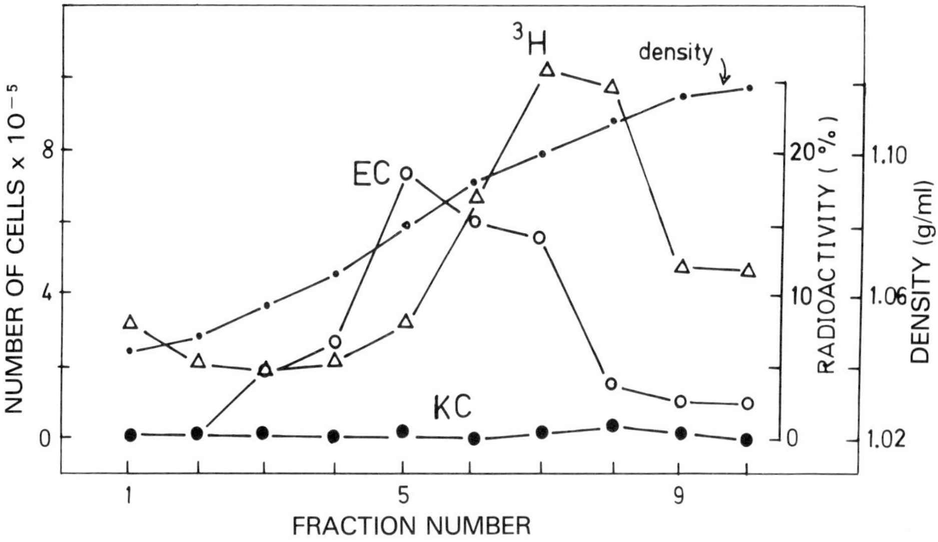

Figure 5. Separation of non-parenchymal rat-liver cells in Nycodenz gradients after treatment with carbonyl iron. Non-parenchymal cells were prepared from a rat given an intravenous injection of carbonyl iron prior to sacrifice (see text). The iron-containing Kupffer cells in the initial cell suspension (obtained by treating the perfused liver with collagenase) were removed by means of a magnet. Non-parenchymal cells were prepared by differential centrifugation and introduced into the gradient as described in the text. Kupffer cells (KC) and endothelial cells (EC) were counted in the fractions. The radioactivity was determined to indicate the distribution of stellate cells.

distribution of retinol-containing (stellate) cells.

2.4.3. *Isolation of Stellate Cells*

Stellate cells have a tendency to adhere to other non-parenchymal cells. This property makes purification of these cells difficult. One approach that has given fairly pure preparations of stellate cells is described in *Table 1* and employs pronase prepared non-parenchymal cells (12). The perfused liver is treated with both collagenase and pronase and the final preparation of non-parenchymal liver cells is separated on a two-step gradient of metrizamide or Nycodenz *(Table 1)*. The stellate cells accumulate at the top of the upper layer of the gradient which contains 13% (w/v) Nycodenz (or metrizamide). The stellate cells can be further purified by centrifugal elutriation. After purification the stellate cells attach to tissue culture dishes, divide and can be kept in culture for many days.

2.5. **Conclusions**

Obviously, the methods for separating the various types of liver cells still need some improvement. The data presented here, show, in accordance with earlier work (15), that partial separation of hepatocytes and non-parenchymal cells is possible in gradients of iodinated gradient media. The distributions of non-parenchymal cells in density gradients of Nycodenz also show partial separations of Kupffer cells, endothelial cells and stellate cells. These preliminary data are promising but, to optimise the separation of these cells, the influences of several parameters need, to be tested, such as changes in the osmolarity, pH of the medium, temperature, shape of the gradient and the cell concentration loaded onto each gradient. However, even using the present techniques as described here, the density distribution of liver cells in Nycodenz gradients can be used for analytical purposes, that is to determine the cellular location of components such as retinol or endocytosed proteins by comparing their density distribution with that of a known cell marker such as peroxidase activity. It

Table 1. Preparation of Stellate Cells (53).

1.	Prepare the non-parenchymal cells as described in Section 2.3.2.
2.	Suspend the non-parenchymal cells in Gey's balanced salt solution (7.00 g NaCl, 0.37 g KCl, 0.17 g $CaCl_2$, 0.07 g $MgSO_4.7H_2O$, 0.21 g $MgCl_2.6H_2O$, 0.15 g $Na_2HPO_4.2H_2O$, 0.03 g KH_2PO_4, 1.00 g glucose, 2.27 g $NaHCO_3$ per litre) gassed with 5% CO_2 in air.
3.	Prepare a solution of 30% (w/v) metrizamide in a salt solution containing 0.17 g $CaCl_2$, 0.07 g $MgSO_4$, 0.21 g $MgCl_2.6H_2O$, 0.15 g $Na_2HPO_4.2H_2O$, 0.03 g KH_2PO_4, 1.00 g glucose and 2.27 g $NaHCO_3$ per litre.
4.	Mix 4 ml of Gey's balanced salt solution with 6 ml of 30% metrizamide solution and pipette into a centrifuge tube.
5.	Mix 10.5 ml of the cell suspension with 8 ml of metrizamide solution and carefully layer this solution over the denser solution in the centrifuge tube.
6.	Centrifuge the discontinuous gradient at 1400xg for 17 min at 15°C.
7.	The upper layer of the gradient is enriched in stellate cells although some Kupffer and endothelial cells do contaminate this fraction.
8.	If necessary the stellate cells can be purified further by elutriation (53).

should soon be possible to utilise modifications of the analytical procedure for preparative purposes also.

3. ISOLATION AND FRACTIONATION OF BLOOD CELLS

3.1. Introduction

Efficient separation procedures are required in order to study the function of the different cell types in blood. For white blood cells distinct density differences have provided a basis for separation of the major subgroups. Mononuclear cells (monocytes and lymphocytes) have a lower density than granulocytes and erythrocytes. By layering the blood over a separation fluid of intermediate density, it is possible to separate the blood cells into two fractions by centrifugation. For these separations compounds originally developed as iodinated X-ray contrast media are usually used as the gradient material. After centrifugation, the mononuclear cells float on top of the separation fluid, whereas the other cells sediment to the bottom of the tube (22–25). In a second step, granulocytes can be separated from the erythrocytes by washing and dextran sedimentation at unit gravity. It is even possible to do this separation using a one step procedure (26,27). Several techniques have been developed for the isolation of blood lymphocytes (28,29). Most of these procedures are based on the removal of unwanted cells (granulocytes and monocytes) by adherence. One problem with such techniques based on adherence is, however, that there may be selective loss of the B-lymphocytes.

Monocytes have a lower density than other cells in human blood and may be separated from other types of blood cell by density gradient centrifugation (30–37). The average monocyte purity is reported to vary from 60–90%, with lymphocytes as the main contaminant. The variability in purity is due to the overlapping densities of these two cell populations (38), and hence it would appear difficult to design an efficient separation procedure based on density differences alone. It is likely that better purity might be obtained by combining different separation techniques. For example, monocytes, partially purified by centrifugation, adhere readily to glass or plastic surfaces. Hence, after incubation at 37°C, non-adhering cells can be removed by washing, and then the monocytes can be detached and recovered by various procedures (39–47).

As demonstrated in previous reports (48,49), the monocyte purity can be improved by using a hyperosmotic density gradient medium. When the osmolarity increases, the cells expel water and shrink. The density of the cells then increases, they move faster and may even pass the density barrier that was present initially. These events can only take place if the cell membrane is semipermeable, allowing free passage of water, whereas there is no or only negligible (net) transport of solutes across the cell membrane. The basis for using these principles for cell separation is that the various cell types have different osmotic sensitivities as a result of, for instance, a different water content. In hypertonic solutions a cell with a high water content will increase its density more than a cell with a low water content. However, an apparently different osmotic sensitivity could also be due to a delayed passage of water through the

Table 2. The Calculated Effects of Increased Osmolarity on the Density, Volume and Radius of Cells.

Calculated effects on cells[a]	*Percentage increase of osmolarity*			
	0[b]	3	10	50
Density (g/ml)	1.0620	1.0632	1.0662	1.0808
Percentage decrease in cell volume	0	2.0	6.4	24.0
Percentage decrease of cell radius	0	0.7	2.2	8.5

[a]70% water in the cell.
[b]Iso-osmolar solution.

cell membrane.

The theoretical effect of increased osmolarity on cell volume, radius and density is shown in *Table 2*. In this part of the chapter the use of the manipulation of the osmolarity and density of solutions for monocyte separations is demonstrated. It appears that the lymphocytes, which overlap in density with monocytes, are more sensitive to increases of osmolarity than are monocytes. A new, nonionic iodinated gradient medium, Nycodenz, has been used in these separation studies, and sodium chloride has been used to adjust the osmolarity of solutions. With the correct combination of density and osmolarity all of the cells loaded onto the gradient, except for the low-density fraction of the monocytes, move to the bottom during centrifugation. A monocyte fraction of 95–98% purity can then be collected from the top of the gradient layer. This gradient medium has also been used for separation of mononuclear cells and granulocytes, in comparison with already established procedures.

3.2. Gradient Media for the Separation of Blood Cells

Sodium metrizoate[1] solution (32.8%, density 1.200 g/ml), Lymphoprep (9.6% sodium metrizoate, 5.6% Ficoll), Lymphopaque (11.6% sodium metrizoate, 4.2% dextran) and Nycodenz, trade name for 5-(N-2,3-dihydroxypropyl-acetamido)-2,4,6-tri-iodo-N,N′-bis(2,3-dihydroxypropyl)isophthalamide, can be obtained from Nyegaard & Co., Oslo, Norway. The 27.6% Nycodenz stock solution has a density of 1.1466 g/ml and osmolarity of 290 mOsm. It is advantageous that by mixing Nycodenz with NaCl solutions of varying concentration, the final density and osmolarity can be adjusted independently of each other over a wide range of densities. To obtain a separation fluid with a chosen density, V parts of Nycodenz are mixed with V_1 parts of a NaCl solution, V_1 being calculated from the following equation:

$$V \times d + V_1 \times d_1 = (V + V_1)d_2 \qquad \text{Equation 1}$$

where d is the density of Nycodenz (1.1466 g/ml), d_1 the density of the NaCl solution, and d_2 the density of the mixture. The greatest accuracy is obtained by using weighed fractions of the two components rather than by measuring volumes. Still it is often necessary to adjust the density of the mixture before use. The desired osmolarity is obtained by varying the concentration of NaCl.

[1]This compound is also available under the trade name of Isopaque.

An alternative procedure for preparing solutions of defined osmolarity is to prepare two solutions of the same density, and high and low osmolarities respectively. By mixing these two solutions in appropriate proportions, the required osmolarity can be obtained. Similarly other separation fluids can be made by combining Nycodenz with Ficoll (8% or more), dextran (6% or more) or sucrose (9.5%). The high molecular weight compounds (Ficoll and dextran) are then dissolved in a NaCl solution of appropriate concentration to achieve the desired osmolarity.

An isotonic Percoll (polyvinylpyrrolidone-coated silica gel particles, Pharmacia, Uppsala, Sweden) separation fluid can be prepared by mixing nine volumes of the stock solution with one volume of 9% NaCl. The density is measured and the suspension is further diluted with 0.9% NaCl to desired density as described for Nycodenz (Equation 1).

3.2.1. *Measurement of Density*

The density of solutions can be measured using a 10 ml or 5 ml pycnometer. It is important to check the accuracy of the pycnometer by weighing it with and without water, and to make allowance for the temperature. Alternatively determine the density electronically with a densitometer (DMA 40, Anton Paar, Graz, Austria), based on the variation of the natural frequency of a hollow oscillator, when filled with liquids of varying densities.

3.2.2. *Measurement of Osmolarity*

The osmolarity of solutions can be determined from freezing point depression measurements using an osmometer (e.g. Advanced Instruments Inc., Needham Heights, MA 02194, USA.).

3.3. **Collection and Preparation of Blood Cells**

Collect human blood in 12 ml Vacutainers containing 14 mg disodium EDTA. Colloidal iron particles in suspension (Lymphocyte Separating Reagent) can be obtained from Technicon Instruments Corp., Tarrytown, NY, USA.

3.3.1. *Defibrination of Blood*

Defibrinate the blood to remove platelets by shaking 10 ml of blood in a tube containing 20–25 glass beads (2–3 mm diam.). Larger volumes can be defibrinated by manual swirling of 100 ml flasks containing 50 ml of blood and 15 glass beads (5 mm diam.) for 10 min.

3.3.2. *Removal of Red Blood Cells by Dextran Sedimentation*

Dissolve 6.0 g of dextran 500 (Pharmacia, Uppsala, Sweden) in 100 ml of 0.9% NaCl. Mix ten volumes of anti-coagulated blood with one volume of dextran solution in a tube to give a liquid column 50–80 mm high. Remove the plasma layer containing the leucocytes when the red cells have settled, usually after about 15–40 min.

3.4. Analysis and Treatment of Gradient Fractions

3.4.1. *Removal of Erythrocytes by NH_4Cl-lysis*

After centrifugation, resuspend the pelleted cells (granulocytes and erythrocytes) in 3 ml of 0.83% NH_4Cl, 10 mM Hepes-NaOH buffer (pH 7) and incubate the cell suspension for 7 min at 37°C.

3.4.2. *Non-specific Esterase Staining for Identification of Monocytes*

This is done as described elsewhere (50,51). Smear a drop of the cell suspension onto a clean microscope slide and allow it to dry. Fix the air-dried preparations for 30 sec in cold (4°C) formalin-acetone (30 ml distilled water, 45 ml acetone, 25 ml 40% formaldehyde, buffered with phosphate to pH 6.6). Wash the slide three times in water and incubate it for 45 min at room temperature in the following medium: 40 mg α-naphthyl butyrate (Sigma, St Louis, MO, USA) in 2 ml ethylene glycol monomethyl (E.Merck, Darmstadt, Germany), 38 ml 0.1 M sodium phosphate buffer (pH 6.3), 0.1 ml 4% pararosaniline (Sigma) in 2 M HCl, 0.1 ml 4% sodium nitrite (Merck) in distilled water. After incubation wash the slides in water and counterstain with Giemsa (1:10) for 6 min. Enzymatic activity is shown by dark red staining of the cytoplasm of the monocytes and macrophages.

3.5. Separation of White Cells from Human Blood

3.5.1. *Separation of Mononuclear Cells and Granulocytes*

Mononuclear cells (monocytes and lymphocytes) have a lower density than erythrocytes and granulocytes. Thus, after centrifugation of blood cells loaded on a separation fluid of appropriate density (1.077 g/ml), the mononuclear cells remaining at the top are easily collected. The granulocytes in the bottom fraction may next be separated from the erythrocytes by dextran sedimentation. All centrifugation steps are carried out at room temperature.

(i) *Separation of cells from whole blood.* This is done as follows: mix equal parts of anticoagulated (or defibrinated) blood and 0.9% NaCl and layer up to 6 ml of the diluted blood over 3 ml of the separation fluid *(Figure 6)*. Whilst layering keep the pipette (10 ml) against the tube wall 15–40 mm above the fluid meniscus. To minimise mixing of the two layers the blood should flow out of the pipette continuously. After centrifugation for 20 min at 600xg (1900 rev/min), the blood cells will have separated into two fractions, a white layer consisting of mononuclear cells and platelets at the interface region, and a pelleted fraction containing the erythrocytes and granulocytes. First remove the supernatant down to 3–4 mm above the white band. Then collect the mononuclear cells using a Pasteur pipette together with the top half of the separation fluid. The cells seem to be located mostly around the periphery of the tube, but to ensure complete removal of cells move the pipette over the whole cross-sectional area. Thereafter remove the rest of the separation fluid down to 2 mm above the erythrocyte mass. Resuspend the cells in the pellet in 1.5–2.0 ml of 0.9% NaCl, transfer to another tube and add an additional

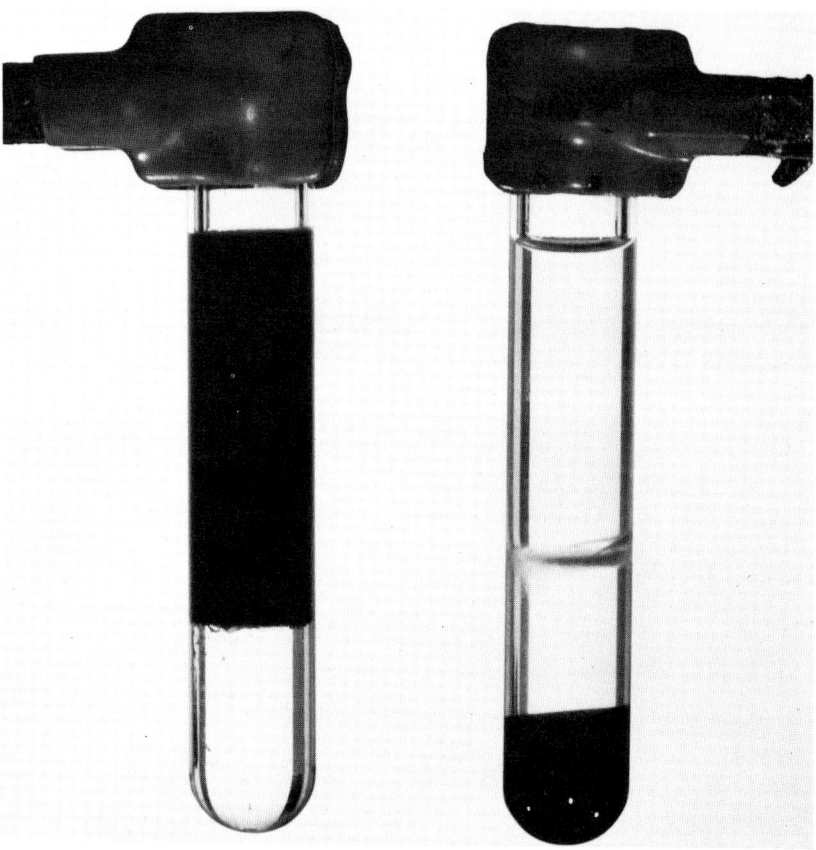

Figure 6. The metrizoate-Ficoll separation technique. Equal parts of anticoagulated or defibrinated blood and 0.9% NaCl are mixed, and 6 – 8 ml of this mixture are layered over 3 ml of metrizoate-Ficoll in a centrifuge tube with inner diameter of 12 – 14 mm. After centrifugation at room temperature for 20 min at 600xg (1900 rev/min), mononuclear cells (and platelets) are found as a white band at the top of the separation fluid, while erythrocytes and granulocytes form a pellet at the bottom of the tube. The mononuclear cells are collected and cells at the bottom are suspended in 6 ml of 0.9% NaCl and centrifuged (7 min at 500xg). The supernatant is removed, and the original blood volume (3 ml) is restored by the addition of 0.9% NaCl and 0.3 ml of 6% dextran is then added. The supernatant containing the granulocytes is harvested when the red cells have settled.

volume of 5 ml of 0.9% NaCl prior to centrifugation for 7 min at 600xg (1900 rev/min). Remove the supernatant and make it up to the original volume (3 ml) by the addition of 0.9% NaCl (or medium). At this stage the cells from 2 – 3 tubes (6 – 9 ml) can be combined in one 12 ml tube, next add 10% (by volume) of 6% dextran. After sedimentation of red cells, collect the supernatant containing the granulocytes. As demonstrated in *Table 3*, the four different separation fluids containing Nycodenz all yielded pure suspensions of mononuclear cells, comparable to those obtained using established procedures with metrizoate-Ficoll and Percoll. In the case of the granulocyte fractions, the lowest purity was observed using the Nycodenz-sucrose solution.

Table 3. Separation of Mononuclear Cells with Different Gradient Media. Equal parts of EDTA-treated blood and 0.9% NaCl were mixed and 4 ml of this mixture were layered on 3 ml of the separation fluid and centrifuged for 20 min at 600xg (1900 rev/min). The density of the metrizoate-dextran solution was 1.085 g/ml, the other separation fluids had a density of 1.077 g/ml. The Nycodenz solutions were prepared by mixing the Nycodenz stock solution with 0.9% NaCl, 8% Ficoll (in 0.95% NaCl), 6% dextran (in 0.95% NaCl) and 9.5% sucrose. T: top fraction. B: bottom fraction. The mean values (± S.E.) are from 6 experiments.

Medium	*Osmolarity (mOsm)*	*Fraction*	*Differential counts (%)*			*Percentage erythrocyte contamination*
			Granulocytes	*Lymphocytes*	*Monocytes*	
Nycodenz-NaCl	314	T		78 ± 3	22 ± 3	10 ± 7
		B	98.3 ± 0.5	1.3 ± 0.2		
Nycodenz-Ficoll	325	T		81 ± 2	19 ± 1	10 ± 7
		B	98 ± 0.3	1.4 ± 0.2		
Nycodenz-dextran	320	T		80 ± 3	20 ± 3	6 ± 1
		B	97 ± 0.6	2.4 ± 0.6		
Nycodenz-sucrose	317	T		86 ± 3	13 ± 2	15 ± 4
		B	95 ± 0.8	3.8 ± 0.6		
Metrizoate-Ficoll[a]	312	T		82 ± 2	18 ± 2	7 ± 3
		B	99.6 ± 0.2	0.4 ± 0.2		
Metrizoate-dextran[b]	348	T		81 ± 3	19 ± 3	12 ± 2
		B	98 ± 0.6	2 ± 0.5		
Percoll	325	T		81 ± 3	19 ± 3	12 ± 6
		B	99.7 ± 0.1	0.3 ± 0.1		

[a]Lymphoprep [b]Lymphopaque

From these results it appears that the Nycodenz-NaCl solution is ideal for routine separation procedures.

(ii) *Separation of cells after removal of the red blood cells by dextran sedimentation.* For separation of cells from large blood volumes (>50 ml), it is preferable to remove erythrocytes by dextran sedimentation in an initial step, although this entails about a 30% loss of leucocytes. Aggregate the red blood cells by adding a tenth volume of 6% dextran solution as described in Section 3.3.2. The height of the blood column should be 50 – 80 mm. The sedimentation of red blood cells is independent of the width of the tube or beaker, if the diameter is 12 – 13 mm or more. Collect the leucocyte-rich plasma when the red cells have settled after 15 – 40 min, and layer (2 – 7 ml) over 3 ml of the separation fluid. After centrifugation, the mononuclear cells are found at the interface region. The granulocytes at the bottom are contaminated by 2 – 5 erythrocytes per granulocyte. If required, the red cells can be removed by NH_4Cl-lysis (see Section 3.4.1). With citrated blood the four different separation fluids gave approximately similar purity and yields of mononuclear cells *(Table 4)*. Similar results were obtained with EDTA-treated blood. In other experiments it was found that metrizoate-Ficoll and Percoll solutions tended (borderline significance) to yield slightly more pure granulocyte suspensions than metrizoate-dextran and Nycodenz-NaCl solutions; this was even more pronounced when using citrated blood from the blood bank and stored 3 – 5 h after withdrawal. The granulocyte yield did not differ significantly, but still it

Table 4. Separation of Mononuclear Cells and Granulocytes from Leucocyte-rich Plasma. Ten volumes of citrated blood were mixed with one volume of 6% dextran (in 0.9% NaCl). After sedimentation of the red cells, 2 ml of the leucocyte-rich plasma were layered over 3 ml of the separation fluid and centrifuged at 22 – 24°C for 20 min at 600xg (1900 rev/min). Then mononuclear cells were harvested from the interface region, and granulocytes from the bottom of the tube. T: top fraction. B: bottom fraction. The cell yield is expressed as the number of cells recovered as a percentage of the total number of cells layered on the separation fluid. The mean values (± S.E.) are from 5 separations.

Medium	*Fraction*	*Differential counts*			*Cell yield*	
		Granulocytes	*Lymphocytes*	*Monocytes*	*Granulocytes*	*Mononuclear cells*
Nycodenz-NaCl	T	0.1 ± 0.1	77.7 ± 3.8	21.8 ± 3.8		88 ± 9
	B	96.7 ± 1.1	2.2 ± 0.6	1.0 ± 0.7	64 ± 9	
Metrizoate-dextran[a]	T	0.7 ± 0.2	76.4 ± 2.5	22.3 ± 2.2		94 ± 16
	B	96.7 ± 0.4	3.0 ± 0.6	0.3 ± 0.2	64 ± 10	
Metrizoate-Ficoll[b]	T	1.3 ± 0.9	75.3 ± 3.7	22.7 ± 4.2		97 ± 16
	B	99.7 ± 0.2	0.3 ± 0.2		57 ± 7	
Percoll	T	1.3 ± 0.7	70.4 ± 3.9	28.1 ± 4.4		105 ± 14
	B	99.0 ± 0.4	1.0 ± 0.4		57 ± 9	

[a]Lymphopaque [b]Lymphoprep

Table 5. Removal of Monocytes from Leucocyte Fractions by Preincubation with Colloidal Iron Particles. Leucocytes from whole blood obtained by centrifugation were incubated with colloidal iron particles and then separated with metrizoate-Ficoll in parallel with a control group of samples in which 4 ml of a 1:1 mixture of heparinised blood and 0.9% NaCl were loaded on the separation medium. The lymphocyte yield is calculated as the number of cells recovered in percent of lymphocytes in unseparated (uncentrifuged) blood. The mean values (±S.E.) are from 6 separations.

	Differential counts				
	Granulocytes	*Lymphocytes*	*Monocytes*	*Basophils*	*Lymphocyte yield*
Control	0.1 ± 0.04	78.3 ± 3.8	20.9 ± 3.8	0.7 ± 0.2	106 ± 8
Incubated	2.6 ± 1.2	96.1 ± 1.2	1.0 ± 0.2	0.3 ± 0.1	55 ± 6

Table 6. Recipes for Nycodenz-NaCl Separation Fluids. This table gives the grams of a NaCl solution to be mixed with 10 ml (11.47 g) of 27.6% Nycodenz in order to obtain a separation fluid of the desired density and osmolarity.

Percentage concentration of NaCl solution	*Grams of NaCl solution*	*Final density (g/ml)*	*Final osmolarity (mOsm)*
0.85	14.51	1.062	300
0.91	14.65	1.062	310
1.02	12.65	1.068	330
1.08	12.75	1.068	340
0.90	9.57	1.077	310
1.40	10.10	1.078	390
1.47	10.14	1.078	400

was strikingly low in some experiments (5 of 20) with Percoll, possibly due to loss by adherence. In preparing the Percoll separation fluid, great care should be taken to avoid bacterial contamination, since Percoll cannot be autoclaved after it has been diluted with NaCl to give an isotonic solution.

3.5.2. *Separation of Lymphocytes*

Lymphocytes and monocytes co-purify when blood is fractionated using the metrizoate-Ficoll method (*Tables 3* and *4*). By capitalising on the ability of monocytes to engulf colloidal iron particles, their density can be increased sufficiently to enable them to sediment through the gradient medium, whereas lymphocytes remain at lower densities.

Collect 10 ml of venous blood in Vacutainers containing heparin. Centrifuge the tubes for 10 min at 600xg (1900 rev/min). Remove the leucocyte layer ($\sim$1 ml) resting on the top of the erythrocyte pellet. Mix the leucocytes with 1 ml of the colloidal iron suspension and incubate the mixture at 37°C in a shaking bath for 30 min. If needed, cells from several 10 ml tubes can be mixed but the height of the mixture should not exceed 10 mm during incuba-

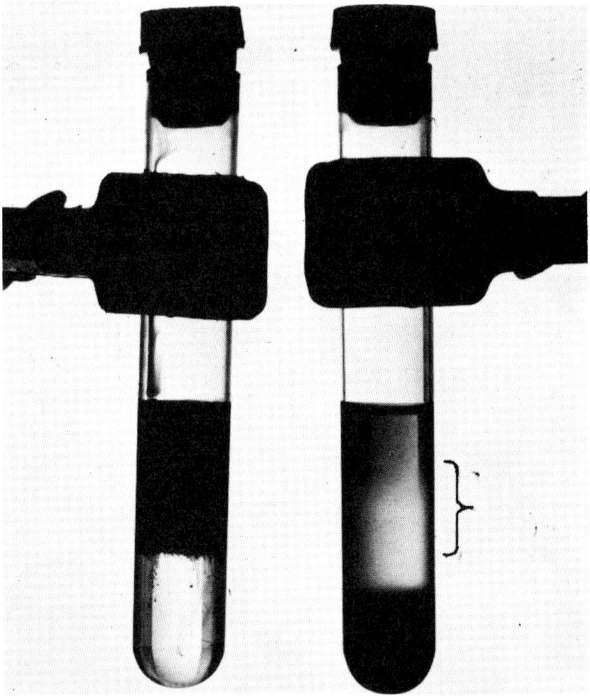

Figure 7. Illustration of the monocyte separation technique. A sample of 3 ml of EDTA-treated blood (as shown), or 2–6 ml of leucocyte-rich plasma are layered over 3 ml of the separation fluid, and centrifuged at room temperature for 15 min at 600xg (1900 rev/min). After centrifugation the clear plasma is removed down to 3–4 mm above the interface. Thereafter, as indicated by the bracket, the remaining plasma together with slightly more than half the volume of the separation fluid are collected. The cells are counted and smears made for differential counting.

tion. After 30 min mix the cell suspension with an equal volume of 0.9% NaCl and layer 4–6 ml of the mixture over 3 ml of metrizoate-Ficoll solution and centrifuge for 20 min at 600xg (1900 rev/min). As shown in *Table 5*, after centrifugation an almost pure suspension of lymphocytes is obtained from the interface region.

3.5.3. *Factors Affecting the Purification of Monocytes*

Monocytes have a lower average density than lymphocytes but, because the densities of the two types of cell overlap, it is difficult to establish a satisfactory separation of high reproducibility, based on density differences alone. However, the separation of monocytes and lymphocytes can be improved by increasing the osmolarity of the gradient medium. The cells then expel water, shrink and their density increases. In this respect the lymphocytes are more sensitive than monocytes, and thus they sediment further during centrifugation, whereas the monocytes remain at the top of the gradient. The practical application of these principles is best understood by following, in a step by step manner, how the procedure was developed. The recipes for the preparation of the various Nycodenz-NaCl solutions used are given in *Table 6*.

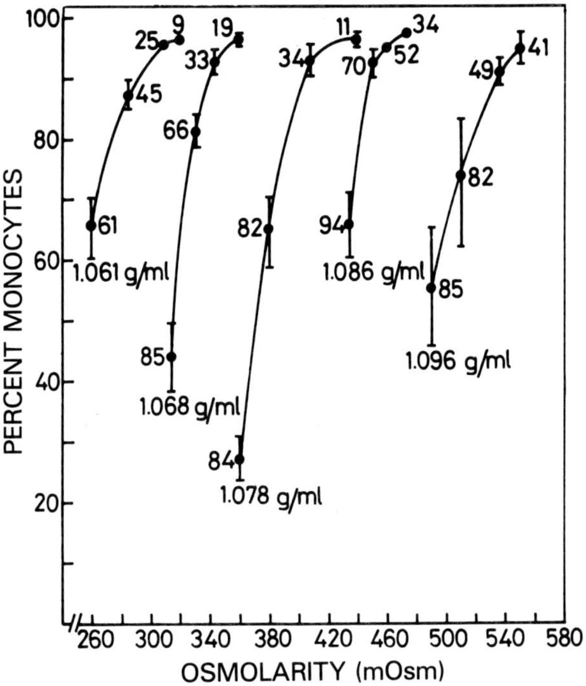

Figure 8. Monocyte separation using Nycodenz-NaCl separation fluids of varying density and osmolarity. A sample of 1.5 ml of leucocyte-rich plasma obtained by dextran sedimentaion was layered over 3 ml of the separation fluid and centrifuged. The yield of monocytes is indicated at each experimental point, which represents the mean (±S.E.) from 5 separations which were not run in parallel.

(i) *Isolation of monocytes from leucocyte-rich plasma using Nycodenz-NaCl solutions of varying osmolarity and density.* After dextran-sedimentation of EDTA-treated blood (section 3.3.2), layer 1.5 ml of the leucocyte-rich plasma over 3 ml of the Nycodenz-NaCl solution and centrifuge for 15 min at 600xg (1900 rev/min). There is no distinct band at the interface after centrifugation, but the separation fluid itself has a greyish colour *(Figure 7)*, mostly caused by non-pelleted platelets. Firstly, remove the clear plasma down to 3 – 4 mm above the interface between plasma and the separation fluid. Next collect the remaining plasma together with slightly more than half the volume of the separation fluid and count the cells before making smears for microscopic examination. Erythrocytes and granulocytes sediment to the bottom during centrifugation, and the cells in the upper half of the Nycodenz solution are almost exclusively lymphocytes and monocytes. The results with Nycodenz solutions of five different densities ranging from 1.06 – 1.096 g/ml are shown in *Figure 8*. Starting with the lowest density (1.061 g/ml), it is clear that as the osmolarity increases, the monocyte purity improves, up to a maximum of 97%. The results obtained were similar at each density, but as the density increased, it was necessary to adjust the osmolarity to a higher level. However, for routine use it appears reasonable to choose a separation in the lower density region. *Figure 9* shows monocytes isolated by this procedure more than 95% of the cells appeared to be monocytes as shown by positive esterase staining (Section 3.4.2).

In a search for the most efficient procedure for the purification of

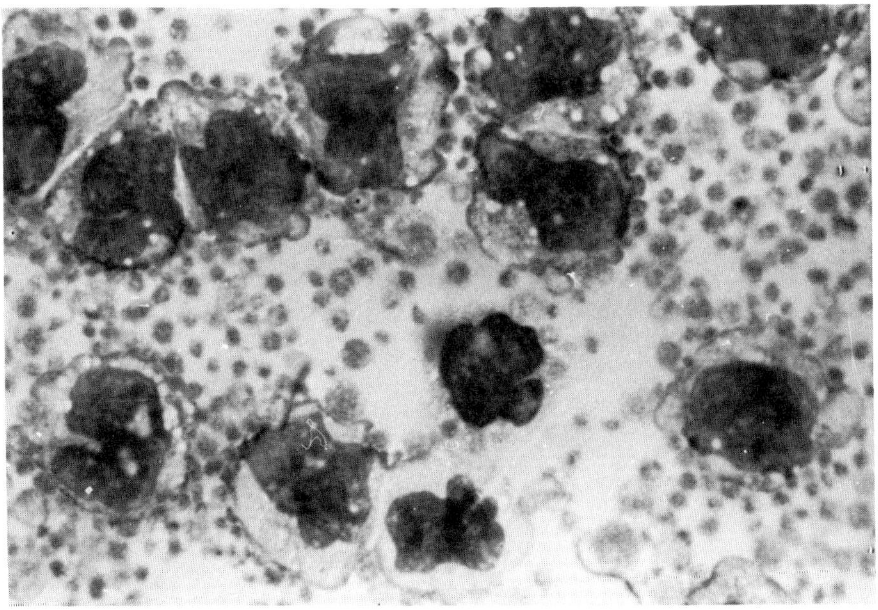

Figure 9. The appearance of purified monocytes separated using Nycodenz-NaCl solution. Magnification x 1000.

monocytes, a more detailed study was done in which the osmolarity was increased stepwise by only 2.5 – 4%, which theoretically would increase the density of the cells by 0.001 – 0.002 g/ml. As shown in *Table 7* this was sufficient to cause a striking improvement in the purity of the monocytes at each density tested. Generally it appears that the low density fractions of the lymphocytes are somewhat more sensitive to an increase in osmolarity than are the monocytes. This implies that when exposed to hyperosmotic Nycodenz solutions, the lymphocytes expel water, and their density increases sufficiently to enable them to sediment through the separation fluid. However, the

Table 7. The Effect of Density and Osmolarity of Nycodenz-NaCl Solutions on the Purity and Yield of Monocytes. Following dextran sedimentation of erythrocytes, 1.5 ml of the plasma were layered over 3 ml of the Nycodenz-NaCl separation fluid and centrifuged at 22 – 24°C for 15 min at 600xg (1900 rev/min). The cells in the interface region between plasma and the separation fluid were removed with a Pasteur pipette and counted. Separation fluids with densities of 1.068 and 1.078 g/ml were tested in parallel. The mean values ± S.E. and from 5 experiments.

Density (g/ml)	*Osmolarity (mOsm)*	*Differential counts (%)*		*Yield (%)*		*Monocytes recovered/ml of plasma (x10^{-3})*
		Lymphocytes	*Monocytes*	*Lymphocytes*	*Monocytes*	
1.062	268	62 ± 3	33 ± 4	56 ± 7	91 ± 8	583 ± 73
	281	45 ± 3	55 ± 3	20 ± 5	93 ± 7	596 ± 69
	288	17 ± 6	83 ± 6	7 ± 4	76 ± 8	471 ± 40
	299	4 ± 2	96 ± 2	1.2 ± 0.9	60 ± 8	383 ± 64
	308	4 ± 1	96 ± 2	0.5 ± 0.2	33 ± 3	212 ± 29
	325	2 ± 1	98 ± 2	0.1 ± 0.05	10 ± 2	64 ± 15
1.068	308	37 ± 5	63 ± 5	14 ± 2	88 ± 5	589 ± 42
	322	20 ± 3	80 ± 3	5 ± 1	76 ± 9	501 ± 51
	327	9 ± 3	91 ± 3	1.8 ± 0.6	62 ± 5	411 ± 34
	340	4 ± 1	96 ± 1	0.3 ± 0.04	41 ± 7	266 ± 50
	349	2 ± 1	98 ± 1	0.15 ± 0.06	28 ± 6	193 ± 42
1.078	369	31 ± 5	69 ± 5	11 ± 3	85 ± 5	561 ± 24
	378	17 ± 4	82 ± 5	5 ± 2	73 ± 3	488 ± 41
	389	7 ± 3	93 ± 3	1.3 ± 0.6	59 ± 7	396 ± 57
	399	6 ± 2	94 ± 2	0.7 ± 0.2	38 ± 5	256 ± 42
	410	3 ± 2	96 ± 2	0.2 ± 0.1	26 ± 6	178 ± 42

Table 8. Variability of the Monocyte Separation Procedure. Leucocyte-rich plasma was obtained by dextran sedimentation, and 1.5 – 2 ml were layered over 3 ml of Nycodenz-NaCl separation fluid (density of 1.062 g/ml), and centrifuged at 22 – 24°Cfor 15 min at 600xg (1900 rev/min). Mean values are given and the ranges are shown in brackets.

Osmolarity (mOsm)	*Number of observations*	*Monocyte percentage*	*Monocyte yield (%)*	*Monocytes recovered/ml plasma (10^{-3})*
300	23	94 (83 – 100)	55 (29 – 98)	285 (84 – 572)
310	17	96 (90.5 – 98.5)	26 (9 – 60)	141 (56 – 318)

monocytes are also affected, and there is a concomitant decrease in the yield of monocytes as the osmolarity and purity of the monocytes increases *(Table 7)*. The variability in several separations, using Nycodenz solutions with density of 1.062 g/ml, is further demonstrated in *Table 8*. Monocytes of high purity are obtained using an osmolarity of 310 mOsm, but the yield varies considerably. This can partly be explained by inevitable errors involved in the differential counting procedure of the original sample because of the low percentage of monocytes among blood leucocytes. The wide range of the yield of monocytes obtained per millilitre of plasma, reflects the variability of monocyte concentration in blood. It should be noted that when the yield is low, the cells recovered are enriched with low-density monocytes; this would

Table 9. Monocyte Separation Using Various Volumes of Unfractionated Leucocytes. Each centrifuge tube contained 3 ml of Nycodenz-NaCl (density = 1.068 g/ml) and either 1.5 ml or 6 ml of leucocyte-rich plasma obtained by dextran sedimentation of EDTA-treated blood. The mean values (±S.E.) are from 3 experiments.

Osmolarity (mOsm)	*Plasma volume (ml)*	*Differential counts (%)*		*Monocyte yield (%)*	*Monocytes recovered (x10^{-3})*
		Lymphocytes	*Monocytes*		
330	1.5	6 ± 2	94 ± 2	73 ± 5	647 ± 104
330	6	6 ± 3	94 ± 3	70 ± 3	2516 ± 409
340	1.5	3 ± 0.5	97 ± 0.5	43 ± 5	402 ± 100
340	6	5 ± 3	95 ± 3	40 ± 7	1401 ± 187

Table 10. Separation of Monocytes from Defibrinated Blood. Following dextran sedimentation of defibrinated blood, 1.5 ml of the plasma were layered over Nycodenz-NaCl solution with a density of 1.061 g/ml and centrifuged. The mean value (± S.E.) are from 5 experiments.

Osmolarity (mOsm)	*Differential counts (%)*		*Monocyte yield (%)*	*Percentage "erythrocyte" contamination*
	Lymphocytes	*Monocytes*		
288	15 ± 5	85 ± 5	78 ± 6	57 ± 10
300	4 ± 1	96 ± 1	39 ± 6	73 ± 4
309	3 ± 2	97 ± 2	17 ± 5	73 ± 1

Table 11. Isolation of Monocytes from Whole Blood. In each case 3 ml of EDTA-treated blood were layered over 3 ml of Nycodenz-NaCl solution and centrifuged for 15 min at 600xg. The mean values are from 5 separations.

Density (g/ml)	*Osmolarity (mOsm)*	*Differential counts (%)*		*Monocyte yield (%)*	*Monocytes recovered/ml of blood (x10^{-3})*
		Lymphocytes	*Monocytes*		
1.068	322	10.5 ± 1.5	89.0 ± 1.5	52 ± 2	265 ± 19
1.068	331	11.0 ± 2.8	89.0 ± 2.8	33 ± 5	183 ± 33
1.068	340	5.0 ± 1.8	95.0 ± 2	17 ± 3	74 ± 14
1.078	388	10.0 ± 2.7	90.0 ± 2.7	56 ± 7	312 ± 57
1.078	402	4.6 ± 1.4	95.4 ± 1.4	35 ± 3	168 ± 39
1.078	411	4.2 ± 1	95.8 ± 1	28 ± 5	146 ± 36
1.078	421	3.5 ± 1	95.5 ± 1	20 ± 4	102 ± 27

apply to any density gradient technique.
(ii) *Separation of monocytes from different volumes of unfractionated leucocytes.* The yields of monocytes from different volumes of leucocyte-rich plasma obtained by dextran-sedimentation and separated using a Nycodenz-NaCl solution with a density of 1.068 g/ml can be analysed as shown in *Table 9*. There was no difference in terms of either purity or yield of monocytes whether 1.5 ml or 6 ml of plasma were separated in a single tube.
(iii) *Separation of monocytes from defibrinated blood.* The purity of monocytes using defibrinated blood (Section 3.3.1) is essentially similar to that obtained with EDTA-treated blood *(Table 10)*. However, the yield of monocytes is reduced by more than 50%, probably as a result of a selective loss of monocytes during defibrination (23). The advantage of defibrination is that the platelets are removed. On the other hand, there is a considerable increase of erythrocyte contamination of the monocytes. It appears that the defibrination procedure causes deformation of some erythrocytes and these tend to cosediment with the mononuclear cells.
(iv) *Isolation of monocytes from whole blood by a one-step procedure.* To do this layer 3 ml of EDTA-treated blood over 3 ml of the separation fluid *(Figure 7)*, and recover the monocytes from the interface after centrifugation. The overall pattern obtained is similar to that using leucocyte-rich plasma with the purity of the monocytes improving as the osmolarity increases *(Table 11)*. The yields appear to be somewhat lower than with leucocyte-rich plasma, but overall recovery of monocytes per millilitre of blood is approximately the same since the loss of cells (approx. 30%) that occurs during dextran sedimentation is avoided. In a series of separations using a separation fluid of 400 mOsm, the purity is found to be somewhat lower (90%) with an overall (n = 13) average value of 92% (range 83 – 99%). Using 410 mOsm Nycodenz solution (n = 10) the reproducibility is found to be better with a mean purity of 96% (range 91.5 – 99%).
(v) *Effects of various anticoagulants on the separation of monocytes.* Monocytes of high purity could only be obtained when EDTA was used as a coagulant. With heparinised blood the purity was markedly reduced, and with citrated blood there was no enrichment of monocytes at all. When citrated plasma was replaced by EDTA-treated plasma, the separation was again satisfactory. There is no obvious explanation for this effect. One possibility is that the passage of osmotically active particles over the cell membrane, and thereby the osmotic sensitivity, may depend upon the anticoagulant used. Interesting in this respect is that the release of proteins from leucocytes, that takes place in citrated plasma, is prevented when the cells are suspended in EDTA-treated plasma (52).

3.5.4. *Removal of Platelets from Monocyte Preparations*

The platelet contamination in the monocyte suspension can be reduced by differential centrifugation. To do this centrifuge 10 ml of anticoagulated blood at 200xg (1100 rev/min) for 10 min. Remove the supernatant which contains the majority (70 – 80%) of the platelets and recentrifuge it at 2000xg for 20 min to obtain platelet-depleted plasma for resuspending the cells in. If required, the procedure may be repeated before the separation of monocytes. Platelets can be removed from leucocyte-rich plasma, obtained by dextran-sedimentation by centrifugation for 10 min at 100xg (700 rev/min) before separation using the Nycodenz separation fluid. Still another possibility is to remove the platelets as a final step, after Nycodenz separation. A washing fluid consisting of 0.9% NaCl containing 5% of EDTA-treated plasma is suitable for this purpose, and centrifugation is carried out at 80xg (500 – 600 rev/min) for 10 min. It is inevitable that some loss of cells occurs during differential centrifugation procedures.

3.5.5. *Viability of Purified Monocytes*

The viability was satisfactory in that more than 95% of the purified monocytes have the ability to engulf 5 – 30 Latex particles each (5 experiments). The phagocytic capacity was not influenced by Nycodenz solutions of high osmolarity up to 640 mOsm (3 experiments). Less than 1% of the cells were stained in the trypan blue viability test.

3.5.6. *Recommended Procedure for the Purification of Monocytes*

It is difficult to devise a single procedure for the isolation of monocytes, but some general guidelines can be given. The monocytes should be separated from EDTA-treated blood directly, or from leucocyte-rich plasma obtained by dextran sedimentation of EDTA-treated blood or defibrinated blood. The recipes for making Nycodenz separation fluids are given in *Table 6*. For leucocyte-rich plasma, solutions with densities of 1.062 g/ml, 1.068 g/ml or 1.078 g/ml are recommended, whereas for whole blood a density of 1.078 g/ml (or 1.068 g/ml) should be chosen. However, it is preferable to have complete control of density and osmolarity from measurements. This is best done by preparing two stock solutions for each density, one with a high and one with a low osmolarity. This may be exemplified as follows for a Nycodenz solution of 1.062 g/ml:

(i) 11.47 g of 27.6% (w/v) Nycodenz solution + 14.27 g of 0.7% NaCl.

(ii) 11.47 g of 27.6% (w/v) Nycodenz solution + 14.80 g of 1% NaCl.

The density is measured and adjusted to 1.062 g/ml, and the osmolarity is then determined. By mixing these two stock solutions, separation fluids of desired osmolarity (300 mOsm or 310 mOsm) can be obtained. Stock solutions with densities of 1.068 g/ml can be prepared in an analogous way by changing the concentrations of the NaCl solution used for diluting Nycodenz *(Table 6)*.

For separations from small blood volumes, layer 3 ml of EDTA-treated blood over 3 ml of the separation fluid. For larger volumes, it is preferable to use dextran sedimentation to remove the red blood cells before centrifugation and in this case 6 ml of the cell-rich plasma can be separated in one 12 ml tube. Even more cells can be separated in a single tube if the cells in the plasma are concentrated by centrifugation before separation with Nycodenz. This separation can be carried out by centrifugation for 15 min at 600xg (1900 rev/min), and the monocytes recovered from the interface region. In case of unsatisfactory separation it may be necessary to adjust the centrifugation conditions with respect to speed, and to check the accuracy of the density and osmolarity of the separation fluid.

3.6. **Conclusions**

Nycodenz is a nonionic iodinated gradient medium that can be used for the separation of white blood cells. The physico-chemical properties of Nycodenz are favourable for separating cells, and by mixing Nycodenz with NaCl solutions of varying concentration, the final density and osmolarity can be varied independently over a wide range. The range of densities and osmolarities is much greater than can be obtained with fully dissociated sodium salts of X-ray contrast media (e.g. sodium metrizoate). Separating freshly drawn blood, Nycodenz-NaCl solutions yield pure suspensions of mononuclear cells (monocytes and lymphocytes) comparable to, or slightly better than, those obtained with established procedures using metrizoate-Ficoll or Percoll as gradient media. Nycodenz also appears to be a suitable gradient medium for the isolation of blood monocytes. Monocytes have a lower average density than lymphocytes, but due to overlapping, efficient separation cannot be obtained based on density differences alone. However, monocytes and lymphocytes react differently to hypertonic conditions and hence, by using hypertonic Nycodenz-NaCl solutions, it is possible to obtain pure monocytes using a one-step procedure.

4. ACKNOWLEDGEMENTS

We wish to thank Lill Naess, Kari Holte, Tony Mette Aamodt and Ellen Nordlie for skillful technical assistance.

5. REFERENCES

1. Rickwood,D. (1978) in *Centrifugal Separations in Molecular and Cell Biology* (Birnie,G.D. and Rickwood,D. Eds) Butterworths, London and Boston. p. 169.
2. Ford,T.C. and Rickwood,D. (1982) *Anal Biochem.*, **124**, 293-298.
3. Seglen,P.O. (1976) in *Methods in Cell Biology* vol. 13 (Prescott,D.M. Ed.) Academic Press Inc., New York. p. 29.
4. Munthe-Kaas,A.C., Berg,T. and Seljelid,R. (1976) *Expl. Cell Res.*, **99**, 146.
5. Wisse,E. and Knook,D.L. (1979) in *Progress in Liver Diseases* (Popper,H. and Schaffner,F. Eds.) Grüne and Stratton, Inc., New York. p. 153.
6. Nilsson,M. and Berg,T. (1977) Biochim. Biophys. Acta, **497**, 171.
7. Mills,D.M. and Zucker-Franklin,D. (1969) *Am. J. Pathol.* **54**, 147.
8. Berg,T. and Boman,D. (1973) *Biochim. Biophys. Acta,* **321**, 585.
9. Wandel,M., Norum,K.R., Berg,T. and Ose,L. (1981) *Scand. J. Gastroenterol.,* **16**, 71.
10. Munthe-Kaas,A.C., Berg,T., Seglen,P.O. and Seljelid,R. (1975) *J. Exp. Med.* **141**, 1.
11. Knook,D.L. and Sleyster,E.C. (1977) in *Kupffer Cells and Other Liver Sinusoidal Cells* (Wisse,E. and Knook,D.L. Eds.) Elsevier/North-Holland Biomedical Press, Amsterdam. p. 273.
12. De Leeuw,A.M., Martindale,J.E. and Knook,D.L. (1982) in *Sinusoidal Liver Cells* (Knook,D.L. and Wisse,E. Eds.) Elsevier Biomedical Press, Amsterdam. p. 139.
13. Berg,T., Tolleshaug,H., Ose,T. and Skjelkvåle,R. (1979) *Kupffer Cell Bull.,* **2**, 21.
14. Tolleshaug,H., Skjelkvåle,R. and Berg.T. (1982) *Infect. Immunity,* **37**, 486.
15. Seglen,P.O. (1976) in *Biological Separations in Iodinated Density Gradient Media*

(Rickwood,D. Ed.), IRL Press Ltd., Oxford and Washington. p. 107.
16. Berg,T., Blomhoff,R. and Norum,K.R. (1982) in *Sinusoidal Liver Cells* (Knook,D.L. and Wisse,E. Eds.) Elsevier Biomedical Press, Amsterdam. p. 37.
17. Blomhoff,R., Holte,K., Naess,L. and Berg,T., manuscript in preparation.
18. Knook,D.L., Blansjaar,N. and Sleyster,E.C. (1977) *Expl. Cell Res.* **109**, 317.
19. Stone,A.B. (1974) *Biochem. J.,* **137**, 117.
20. Blomhoff,R., Helgerud,P., Rasmussen,M., Berg,T. and Norum,K.R. (1982) *Proc. Natl. Acad. Sci. (USA)* **79**, 7326.
21. Smedsrød,B., Eriksson,S., Fraser,J.R.E., Laurent,T.C. and Pertoft,H. (1982) in *Sinusoidal Liver Cells* (Knook,D.L. and Wisse,E. Eds.) Elsevier Biomedical Press, Amsterdam. p. 263.
22. Bøyum,A. (1974) *Tissue Antigens,* **4**, 269.
23. Bøyum,A. (1976) *Scand. J. Immunol.,* **5** (Suppl. 5), 9.
24. Bøyum,A. (1968) *Scand. J. Clin. Lab. Invest.,* **21** (Suppl. 97), 77.
25. Kurnick,J.T., Østberg,L., Stegagno,M., Kimura,A.K., Ørn,A. and Sjøberg,O. (1979) *Scand. J. Immunol.,* **10**, 536.
26. English,D. and Andersen,B.R. (1974) *J. Immunol. Meth.,* **5**, 249.
27. Giudialli,J., Philip,P.J.M., Delque,P. and Sudaka,P. (1982) *J. Immunol. Meth.,* **54**, 43.
28. Bøyum,A. (1977) *Lymphology,* **10**, 71.
29. Bøyum,A. (1982) *HLA Typing* (Ferrone,S. and Solheim,B.G., Eds.) CRC Press Inc. Boca Raton, Florida, 1, 1.
30. Bennet,W.E. and Cohn,Z.A. (1966) *J. Exp. Med.,* **123**, 145.
31. Nathanson,S.O., Zamfisrescu,P.L., Drew,S.I. and Wilbur,S. (1977) *J. Immunol. Meth.,* **18**, 225.
32. Gmeling-Meyling,F. and Waldeman,T.A. (1980) *J. Immunol. Meth.,* **33**, 1.
33. Pertoft,H., Johnsson,A., Värmegård,B. and Seljelid,R. (1980) *J. Immunol. Meth.,* **33**, 221.
34. Fluks,A.J. (1981) *J. Immunol. Meth.,* **41**, 225.
35. Hardin,J.A. and Downs,J.T. (1981) *J. Immunol. Meth.,* **40**, 1.
36. Berthold,F. (1981) *Blut,* **43**, 367.
37. Brandslund,J., Møller Rasmussen,J., Fisker,D. and Svehag,S.E. (1982) *J. Immunol. Meth.,* **48**, 199.
38. Loos,H., Blok-Shut,B., van Dorn,R., Hoksbergen,R., de la Riviere,A.B. and Meerhof,L. (1976) *Blood,* **48**, 731.
39. Brodersen,M.P. and Burns,C.P. (1973) *Proc. Soc. Exp. Biol. Med.,* **144**, 941.
40. Alderson,E., Birchall,J.P. and Owen,J.J.T. (1976) *J. Immunol. Meth.,* **11**, 297.
41. Rinehart,J.J., Gormus,B.J., Lange,P. and Kaplan,M.E. (1978) *J. Immunol. Meth.,* **23**, 207.
42. Ackerman,S.K. and Douglas,S.D. (1978) *J. Immunol.,* **120**, 1372.
43. Kragballe,K., Ellegaard,J. and Herlin,T. (1980) *Scand. J. Haematol.,* **24**, 399.
44. Mukherij,B. (1980) *J. Immunol. Meth.,* **37**, 233.
45. Treves,A.J., Yagoda,D., Haimowitz,A., Ramu,N., Rachimilewitz,D. and Fuchs,Z. (1980) *J. Immunol.,* **39**, 71.
46. Tice,D.G., Goldberg,J. and Nelson,D.A. (1981) *J. Reticuloendoth. Soc.,* **29**, 459.
47. Zanella,A., Mantovani,A., Mariani,M., Silvani,C., Peri,G. and Tedesco,F. (1981) *J. Immunol. Meth.,* **41**, 279.
48. Bøyum,A. (1982) Hoppe Zeyler's Z. Physiol. Chem., **363**, 999.
49. Bøyum,A. (1983) *Scand. J. Immunol.,* **17**, 429.
50. Yam,L.T., Li,C.Y. and Crosby,W.H. (1971) *Amer. J. Clin. Path.,* **55**, 283.
51. Svennevig,J.L., Løvik,M. and Svaar,H. (1979) *Int. J. Cancer,* **23**, 626.
52. Fagerhol,M.K., Dale,I. and Andersson,T. (1980) *Scand. J. Haemat.,* **24**, 393.
53. Knook,D.L., Seffelaar,A.M. and de Leeuw,A.M. (1982) *Expl. Cell Res.* **139**, 468.

Preparation of Isolated Rat-liver Cells

T. Berg and R. Blomhoff

1. Solutions

Pre-perfusion Buffer

8.0 g NaCl, 0.4 g KCl, 0.06 g $Na_2HPO_4.2H_2O$, 0.047 g KH_2PO_4, 0.20g $MgSO_4.7H_2O$ and 2.05 g $NaHCO_3$ dissolved in H_2O to 1000 ml final volume.

Perfusion Buffer

35 mg collagenase (130–150 units/mg solid) and 100 μl of 1.0 M $CaCl_2$ made up to 50 ml with H_2O. Both the pre-perfusion and perfusion buffers are gassed with 95% O_2/5% CO_2 to pH 7.5 during perfusion.

Incubation Buffer

8.5 g NaCl; 0.4 g KCl; 0.06 g $Na_2HPO_4.2H_2O$; 0.047 g KH_2PO_4; 0.20 g $MgSO_4.7H_2O$; 4.76 g Hepes; 0.29 g $CaCl_2.2H_2O$. Make up to 1000 ml with H_2O and adjust to pH 7.5 with NaOH. The osmolarity of this solution is approximately 298 mOsm.

2. Method

2.1. *Apparatus*

Set up the perfusion apparatus using a system similar to that shown in *Figure 1*. The important features of this apparatus are; firstly; that it is possible to perfuse the tissue with solutions pre-equilibrated to a temperature of 37°C. Secondly, it is important to be able to switch from the pre-perfusion buffer to the perfusion buffer without any air bubbles entering the system, since air bubbles cause air locks and prevent efficient perfusion of the liver. The pump should be capable of a variable flow rate, from 0–50 ml/min.

2.2. *Perfusion*

Expose the whole abdomen of a deeply ether-anaesthetised rat by a transverse cut across the lower abdomen and a longitudinal cut up to and beyond the sternum. Gently move the gut contents to the left side of the rat to expose the inferior vena cava and the vena porta. Pass a ligature around the vena porta about 1cm from its junction with the liver capsule and tie loosely. With a very fine-pointed scissors, cut through the vena porta, about half its circumference, 0.5 cm on the caudal side of the ligature. Insert the cannula into the vena porta to a point 2 mm past the ligature, do not enter the liver capsule. During this

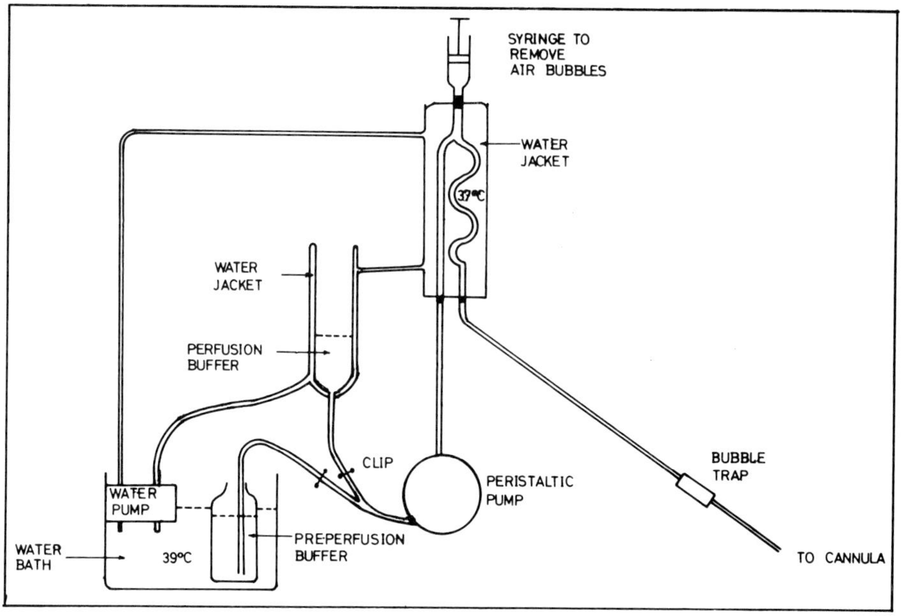

Figure 1. Liver Perfusion Apparatus.

operation the flow rate of the buffer should be reduced to about 10 ml/min. When the cannula is properly in place, tighten the ligature and tie securely. Immediately sever the inferior vena cava to permit a free flow of the pre-perfusion buffer and then increase the flow rate to 50 ml/min. This treatment disrupts the desmosomal cell junctions which are calcium dependent. The rat should now be killed by puncturing the diaphragm. Free the liver by carefully cutting it away from all attachments, taking care not to damage the liver capsule. Continue the pre-perfusion for 10 min with liver *in situ*, ensuring that the lobes lie in their natural position thus ensuring a flow of buffer to all parts of the liver. Then switch to the perfusion buffer and allow 30 sec to elapse before transferring the liver to the perfusion dish so that recirculation of the perfusion buffer can take place. Continue the perfusion for 10 min, by which time it should be possible to see signs of the liver disintegrating within the capsule. Disconnect the cannula and place the liver into a petri dish containing ice-cold incubation buffer. Cut away all extraneous tissue from the liver, taking care not to damage the capsule at this stage. Transfer the liver to another dish of incubation medium and disrupt the capsule, using forceps, and free the cells into the medium by fairly vigorous agitation of the forceps in the medium. Transfer the capsule remnants to fresh medium and free any remaining cells in the same way. Discard the capsule remnants and combine the contents of the two dishes. The total volume of buffer into which the cells are isolated should be about 50 ml. Subsequent treatment of the cell suspension will depend upon the use to which the cells are to be put.

CHAPTER 8

Comparison of Various Density-gradient Media for the Isolation and Characterisation of Animal Viruses

D.A. Vanden Berghe

1. INTRODUCTION

1.1. General Considerations

In virology it is important to have different and effective methods of virus purification in order to study the physico-chemical characteristics of virus particles. Throughout all the various purification steps, it is important to conserve the full infectivity of viruses. It has always been difficult to separate cellular material from viral particles because both consist of similar material. This is especially true for enveloped viruses which are often pleomorphic because they are associated with cellular membranes. They do not have distinct sedimentation coefficients and usually have buoyant densities similar to those of the cellular endoplasmic reticulum. Separation on the basis of filtration or sedimentation is practically impossible for these viruses. Purification and concentration of virions or virus precursors are generally accomplished by fractionation of a culture fluid or a cellular cell extract by various techniques. Freeze-thawing, ultrasonication or grinding techniques can be used and infectivity titrations performed on the separated fractions. Subsequently differential centrifugation, salting-out techniques, isoelectric precipitation, (ultra)-filtration or chromatographic techniques are employed to purify or concentrate the virus particles further. Finally, buoyant density gradient centrifugation is used to characterise the particles. Sometimes viruses can be purified only by sedimentation techniques; this is, for example, the case of some *Picornaviridae*. The method of buoyant density-gradient centrifugation is based on the known differences in buoyant density in a particular gradient medium. The media generally used are sucrose, CsCl, potassium tartrate, metrizamide and sodium metrizoate. The densities of the purified virus particles are used in virus taxonomy as important criteria for identification. Purified virus here means infectious virus, free of cellular material or other viral components. Sometimes differences are noticed between the accepted taxonomic densities and the densities of semi-purified infectious particles because of the association of cellular material with the viruses during buoyant density gradient centrifugation. Therefore the proposed buoyant densities of newly described viruses must be critically evaluated because occasionally there is no indication as to the purity of the virus preparations studied. Virus purification is always monitored by titration of the infectivity. Therefore the purification of viruses needs to be carried out in gradient media which do not influence virus infectivity; similarly, media should not be toxic to cells. Furthermore, inexpensive media are generally preferable because large quantities are often needed and their recovery and reuse is usually difficult, if not impossible. Other favourable properties of media are a lack of interference in the monitoring of gradients; minimal viscosity to shorten the centrifugation time; small enough to be rapidly removed by dialysis and significant differences in the densities of proteins and nucleic acids. Because only a few media are available, some new ones are assessed in this study and the influences of the various established media together with the newly proposed media on the infectivity of the animal RNA viruses *in vitro* and *in vivo* are discussed.

In addition, this study will evaluate the use of density gradient media in virology and the use of virus densities in taxonomy.

1.2. **Properties of Animal RNA Viruses**

The International Committee for Taxonomy of Viruses (1) recognised the family *Picornaviridae* as being divided into four genera. One of the main characteristics of this family is the buoyant density of the viruses in CsCl of 1.33 – 1.45 g/ml depending on the genus. The viruses are spherical naked nucleocapsids, 20 – 30 nm in diameter, they have no envelope or core and infectious particles contain one molecule of positive sense single-stranded RNA (ssRNA). Human poliovirus 1, used in this study as one of the test viruses, is the type species of the genus enterovirus. The main characteristics of this genus are their buoyant densities in CsCl of 1.33 – 1.34 g/ml and a sedimentation coefficient of 150S. Human Coxsackie virus, also used in this study, belongs to the same genus and shows the same density in CsCl. This has been confirmed by various authors (2 – 6) after purification of the viruses by rate-zonal gradient centrifugation.

Polio and Coxsackie viruses are composed of four different polypeptides (6,7). Only these four polypeptides can be found in the infectious CsCl band at 1.33 g/ml after removal of cellular material and virus precursors. The *in vivo* morphogenesis of poliovirus and Coxsackie virus is also very similar. The 5S particles are converted into 14S material which is then converted to 150S infectious material after combination with viral RNA (8). The 74S top components (empty capsids) (9,11) and 45S protein-RNA (10) complexes are also found during virus maturation under certain conditions and depending on the cell system involved. The real precursor of 150S infectious virus is still not known with certainty. Most viral protein synthesis occurs on the endoplasmic reticulum (6,12) which is easily separated from the soluble cytoplasm by low speed centrifugation (12). The virus precursors (5S, 14S, 74S) and 150S virions are liberated into the cytoplasm. After disruption of the infected cell these particles are found extracellularly (6).

Semliki Forest virus belongs to the family of *Togaviridae* (1). These viruses contain a single molecule of positive sense ssRNA but the infectious virus particles have an envelope. There are one or more glycoproteins in the lipoprotein envelope, whose lipids are host cell-derived. The virus particles of the *Togaviridae* are spherical, 40 – 70 nm in diameter with an envelope which is tightly associated with the nucleocapsid (25 – 35 nm diameter) and with buoyant densities of 1.25 g/ml in CsCl and 1.13 – 1.24 g/ml in sucrose gradients. The family is divided into four different genera; Semliki Forest virus belongs to the genus alphavirus. Replication takes place in the cytoplasm and maturation, by budding or pre-assembly of nucleocapsids occurs through the plasma membrane. One of the main physico-chemical properties of the genus alphavirus is their buoyant density in sucrose gradients of 1.21 g/ml and a sedimentation coefficient of 280S. Both poliovirus and Semliki Forest virus are routinely banded in sucrose rate-zonal gradients because they have well-defined sedimentation coefficients. The data in the literature concerning the buoyant densities of alphaviruses are not very consistent and differences in buoyant densities, even in the same medium, have been observed (13 – 15). Comparative studies of buoyant densities of alphaviruses in different gradient media are therefore necessary.

Measles virus is the type species of genus morbillivirus of the family *Paramyxoviridae* (1). The viruses of this family contain one molecule of ssRNA and most particles contain a negative strand. These viruses have a pleomorphic shape, usually roughly spherical, 150 nm or more in diameter. The envelope is derived from the cell surface associated with the viral coded glycoproteins; the nucleocapsids have helical symmetry *(Figure 1)*. The density of these viruses in sucrose gradients is 1.18 – 1.20 g/ml (1) and the viruses have no distinct sedimentation coefficient ($s_{20,w}$ > 1000S). As in the case of Semliki Forest virus, the infectivity of measles virus is destroyed by lipid solvents. It is difficult, if not impossible, to purify measles virus (16,17) by conventional methods; hence the values of the buoyant density of measles virus given in the literature must remain doubtful.

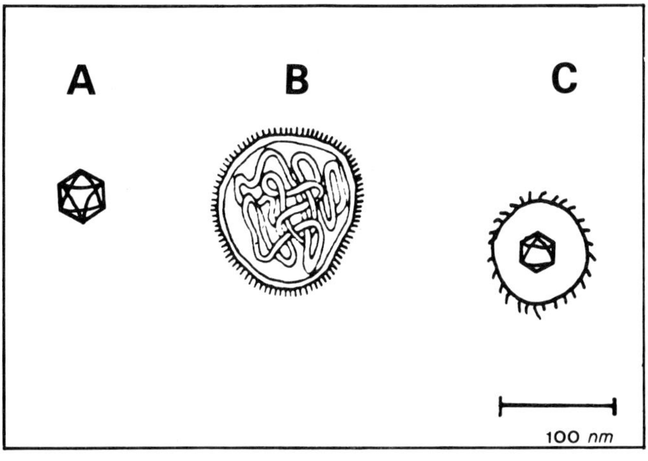

Figure 1. The morphology of various RNA viruses. A) poliovirus and Coxsackie virus; B) measles virus; C) Semliki Forest virus.

This review of the physico-chemical data of the RNA viruses used in this study show that the physico-chemical data of viruses constitute a basis for virus purification and consequently virus taxonomy. Since virus structure and virus stability varies greatly from one virus group to another different techniques must be employed for virus purification and characterisation depending on the density, size and surface properties of the virions.

1.3. Density-gradient Media for the Purification and Characterisation of Viruses

Rate-zonal centrifugation for well-defined and stable particles (e.g., poliovirus, Coxsackie virus and Semliki Forest virus) and equilibrium density sedimentation are frequently used for virus purification; gradient media which are likely to disrupt the integrity of viruses should be avoided. Gradients of caesium salts (chloride, sulphate or formate) are generally used for purifying stable naked viruses, and sucrose (in H_2O or D_2O) gradients for enveloped and the more sensitive viruses. The densities of viruses in CsCl and sucrose gradients are therefore used as taxonomic criteria. Because some viruses lose infectivity in these types of media leading to some disagreement about their buoyant densities, alternative media have been tried for the purification and characterisation of viruses. A series of nonionic "inert" media have become commercially available and have been tested in a number of different laboratories (18). Metrizamide, a nonionic iodinated density gradient medium, is more expensive than sucrose and has been used mainly in virology for the separation of unstable virus particles or very labile viral protein-nucleic acid complexes which could not be resolved or isolated in other types of media. The importance of the introduction of metrizamide can be gauged from its use for the isolation and purification of murine pneumonia virus (19), respiratory syncytial virus (20), channel cat fish virus (21), rubella virus (22), Dane particles containing a DNA strand (23), enveloped mycoplasma viruses (24), different capsids from bacteriophage T-7 (25), H-1 parvovirus subfractions (26), mengovirus (27), fractionation of DNA and membrane-bound DNA and proteins from bacteriophage (28), ribonucleoprotein complexes from influenza virus (29,30), deoxyribonucleoprotein complexes (31), nuclear RNA-protein complexes (32) and core protein precursor proteins (33) from adenovirus type 2 infected cells.

$SrCl_2$ is also considered in this study because it has been observed that buoyant densities in this medium differ from those in CsCl and sucrose gradients. So this gives another physico-chemical characteristic on which to base viral taxonomy. With $SrCl_2$ densities up to 1.34 g/ml are obtainable at 4°C allowing many viruses to be banded by equilibrium centrifugation. Recently Nyegaard and Co (Oslo, Norway) have introduced Nycodenz, a tri-iodinated derivative of benzoic acid, which is nonionic, non-toxic, very water-soluble and autoclavable. In this study the effects of Nycodenz and $SrCl_2$ on virus infectivity has been compared with those of CsCl, sucrose, potassium tartrate and metrizamide.

2. MATERIALS AND METHODS

2.1. Cell Cultures

Stock cultures of Vero cells (25 x 10^6 cells/bottle) are maintained in 1 litre Roux bottles containing Tissue Culture Medium 199 (M 199) (Wellcome, Beckenham, England) and 4% newborn calf serum (Flow Inc., USA). Prior to use, penicillin G (100 U/ml) and streptomycin (0.5 mg/ml) are added to the culture medium. Maintenance medium (MM) used is that described by Hronovsky *et al.* (34) supplemented with 2% newborn calf serum, antibiotics and adjusted to pH 7.2 with 1 M NaOH.

2.2. Test Viruses

Poliomyelitis type 1 strain 1 AS_3 (35) and measles virus strains Edmonston A and B were plaque purified; herpes type 1 was a clinical isolate and identified by immuno-fluorescence and neutralisation reaction. Coxsackie virus strain B2 was obtained from NIH (Bethesda, Maryland, USA); and Semliki Forest L10 was supplied by Dr C.J.Bradish (Porton Down, Salisbury, England). All virus stocks are prepared in Vero cells in maintenance medium. Viral titres of herpes and Semliki Forest viruses are estimated by determining the limiting dilution which can initiate infection in 50% of the cultures inoculated. The 50% endpoints were calculated by the method of Reed and Muench (36). Viral titres of poliovirus, Coxsackie virus and measles virus were estimated by plaque formation on Vero monolayer cells. The viral titres were respectively 10^8 PFU/ml for poliovirus, 10^7 PFU/ml for Coxsackie virus, 10^6 PFU/ml for measles virus, 10^7 TCD_{50}/ml for Semliki Forest and 10^6 TCD_{50}/ml for herpes virus.

2.3. Solutions

Prepare solutions of sucrose, $SrCl_2.6H_2O$, CsCl, potassium tartrate (all from Merck, Darmstadt, GFR), Metrizamide (Nyegaard & Co, Oslo) and Nycodenz (Nyegaard & Co, Oslo) solutions (w/v) in RSB buffer (10 mM Tris-HCl, pH 7.2, 0.1 M NaCl and 1.5 mM $MgCl_2$) or in TNE buffer (10 mM Tris-HCl, pH 7.2, 0.1 M NaCl, 0.02 M EDTA). Titrate the infectivity of viruses in the various incubation media in microtitre plates after serial dilutions as earlier described (37). For *in vivo* studies, 50% (w/v) solutions were diluted with maintenance medium and adjusted with 1 M NaOH to pH 7.2. Sterilise the diluted solutions by filtration through a Millipore membrane filter (0.2 μ) and store at $-30°C$ until used.

2.4. Method for the Equilibrium Centrifugation of Viruses

Clear the supernatants from infected cells of cell debris and centrifuge the supernatants at 50 000xg for 4 h at 5°C to pellet the virus. Take the pellet up in 1 ml of RSB or TNE buffer. Band the virus by equilibrium centrifugation on freshly prepared discontinuous gradients (5 x 2 ml layers) of sucrose, CsCl, metrizamide, potassium tartrate, Nycodenz, $SrCl_2$ in either RSB or TNE buffer in 11 ml tubes for a Beckman SW41 rotor or equivalent rotor. Centrifuge the gradients for 18 h at 200 000xg at 4°C. Fractionate the gradients at 4°C and measure the refractive index of each fraction using an Abbé refractometer. Conversion from refractive indices to densities can be done with the help of reference curves in which refractive indices (at 20°C) are plotted against densities determined at the experimental temperature using high precision pycnometers holding 5.00 ml. Perform infectivity titrations of the fractions in microtitre plates after serial dilution in maintenance medium as described earlier (36).

2.5. Labelling of Viruses

Poliovirus and measles virus are labelled with ^{14}C-leucine, ^{14}C-uridine and L-^{35}S-methionine (Amersham International, Amersham, England) in maintenance medium without calf serum and deficient in the appropriate amino acids as described earlier (12,16). Rate-zonal sucrose gradient centrifugation of labelled poliovirus is carried out as described earlier (12).

3. EFFECTS OF VARIOUS GRADIENT MEDIA ON THE INFECTIVITY OF VIRUSES

3.1. Effects of Density-gradient media on Animal RNA Viruses in Vitro

For all tests the RNA viruses are incubated for 4 h or 18 h at 4°C in RSB buffer containing varying concentrations of density gradient media up to 70% (w/v). Samples are titrated for infectivity and the influence on the virus titre analysed. *Figure 2* shows the results obtained in the case of measles Edmonston A virus and Semliki Forest virus after incubation in various media. Only metrizamide drastically decreases the titre of measles virus; this effect has been reported previously (38). However, in this study (38) all of the virus suspensions were dialysed for 3 h against Tris-HCl buffer. In this study, as mentioned in Section 2.3, samples were diluted in maintenance medium and titrated directly. The decrease in virus titre observed using this method is not so great as when the samples are dialysed. Metrizamide inactivation of measles virus also depends on the strain used (40). For example, Edmonston B virus is not inactivated at all in 30% (w/v) metrizamide. *Table 1* shows the influence

Table 1. Influence of Gradient Media on RNA Viruses *in Vitro*. Viruses were incubated in 50% (w/v) medium for 18 h at 4°C.

	Titre reduction factor					
Viruses	*Potassium tartrate*	*CsCl*	*Sucrose*	*Metrizamide*	*Nycodenz*	*SrCl$_2$*
Coxsackie virus	1.4	2	1	1	1	1.1
Poliovirus	1.3	2.5	1	1	1	1.2
Measles virus	3	6	2	(10^3)	6	30
Semliki Forest virus	3	5	1.3	1.1	1.3	10

( ): depends on the strain used.

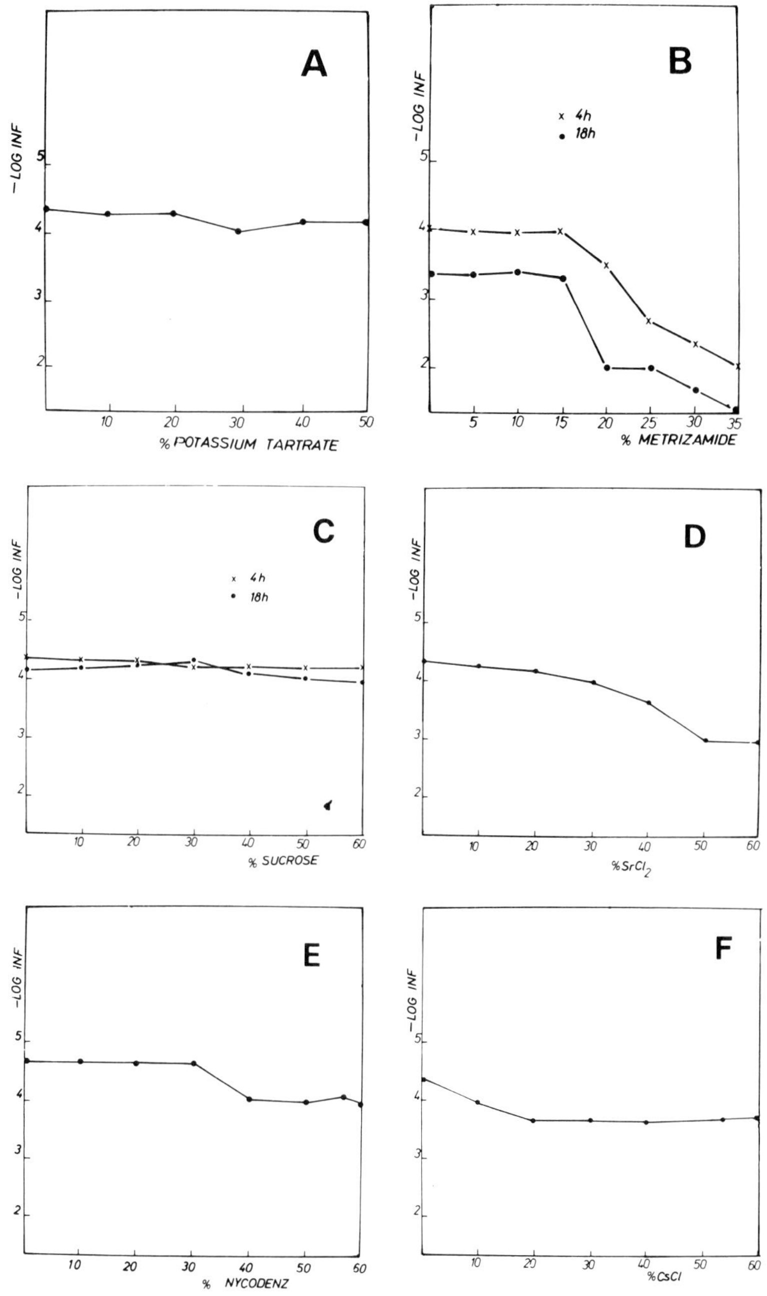

Figure 2. *In vitro* incubation of measles virus Edmonston A in different density gradient media at 4°C. The ordinate shows the log of the infectivity titre and the abcissa shows the concentration of the density gradient medium as percentage (w/v). A) potassium tartrate, 18 h incubation; B) metrizamide, 4 and 18 h incubation; C) sucrose, 4 and 18 h incubation; D) $SrCl_2$, 18 h incubation; E) Nycodenz, 18 h incubation; F) CsCl, 18 h incubation.

Table 2. Influence of Sucrose on the Titre of Viruses *in Vivo*

Titre reduction factor

Viruses	*% (w/v) sucrose in medium*								
	5	*4*	*3*	*2*	*1*	*0.5*	*0.1*	*0.05*	*0.01*
Coxsackie virus	1	1	1	1	1	1	1	1	1
Poliovirus	1	1	1	1	1	1	1	1	1
Herpes virus	1	1	1	1	1	1	1	1	1
Measles virus	1	1	1	1	1	1	1	1	1
Semliki Forest virus	1	1	1	1	1	1	1	1	1

Table 3. Influence of Metrizamide on the Titre of Viruses *in Vivo.*

Titre reduction factor

Viruses	*% (w/v) metrizamide in medium*								
	5	*4*	*3*	*2*	*1*	*0.5*	*0.1*	*0.05*	*0.01*
Coxsackie virus	10^4	10^4	10^3	10^3	10^2	1	1	1	1
Poliovirus	10	10	1	1	1	1	1	1	1
Herpes virus	1	1	1	1	1	1	1	1	1
Measles virus	1	1	1	1	1	1	1	1	1
Semliki Forest virus	10^2	10	10	10	10	1	1	1	1

of 50% (w/v) gradient medium on the infectivity of different RNA viruses after incubation at 4°C for 18 h. The effect on the viral infectivity is expressed by the titre reduction factor. Enveloped viruses are generally affected more than naked viruses. Sucrose, metrizamide and Nycodenz have less effect on both types of virus. The results indicate that all media can be used for equilibrium density-gradient centrifugation.

3.2. Effects of Density-gradient media on Animal RNA Viruses in Vivo

It is important to localise the infectious virus bands after centrifugation as quickly as possible and so the dialysis of the fractions must be omitted. An alternative approach is to study the influence of the media on the virus titre *in vivo*. In all tests the viruses are titrated in microtitre plates. Ninety minutes after virus absorption at 37°C, maintenance medium containing varying concentrations of density-gradient media up to 5% is added to the Vero cells as maintenance medium. Cell and virus controls were included and all experiments were performed in ten-fold. Herpes virus, an enveloped DNA virus, was also included.

3.2.1. *Influence of Sucrose on the Titre of Viruses in Vivo*

Table 2 shows the titre reduction factors after virus titration in sucrose containing maintenance medium. Virus titre is not at all affected by sucrose up to 5% in tissue culture maintenance medium.

3.2.2. *Influence of Metrizamide on the Titre of Viruses in Vivo*

As shown in *Table 3*, the influence of metrizamide depends on the virus used.

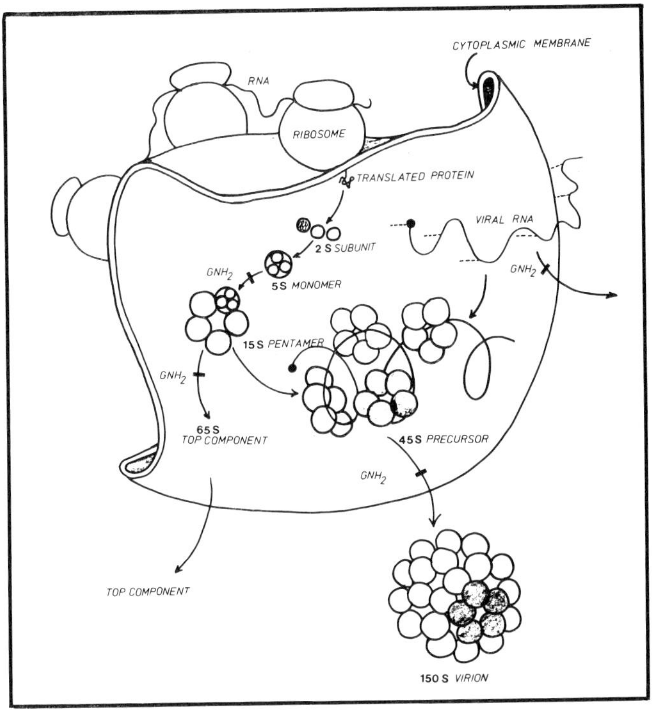

Figure 3. Influence of D-glucosamine (GNH_2) on the morphogenesis of poliovirus on cytoplasmic membranes of Vero cells. GNH_2 block of morphogenesis step.

The titres of all RNA viruses except measles virus were reduced in media containing more than 3% (w/v) metrizamide. Poliovirus and Coxsackie virus, both *Picornaviridae*, are affected to different extents by metrizamide. In 3% metrizamide only 0.1% of the original Coxsackie virus titre remains while the titre of poliovirus is not affected. These results confirm the earlier data of titrations in a HeLa cell system (38) and a mouse organ culture (39) in the presence of metrizamide. The decrease of virus titre in the presence of metrizamide is probably due to the effect of D-glucosamine. Indeed, metrizamide consists of a substituted tri-iodinated benzoic acid nucleus, called metrizoic acid, which is linked by an amide bond to D-glucosamine. Such amide bonds are easily broken *in vivo* so that the D-glucosamine released becomes the active component, since metrizoic acid has no influence at all on the titre of these viruses (40). D-glucosamine affects both enveloped and naked viruses at extra-cellular concentrations greater than 1 mM (41). *Figure 3* shows the influence of glucosamine on the morphogenesis of polio virus in Vero cells. The conversion of 14S particles into 74S top components is blocked. This is the result of a change in the cellular endoplasmic reticulum, the site of the poliovirus biosynthesis, and not due to an effect on the specific polio proteins (10,41). In all cases, the viruses most affected by metrizamide *in vivo*, are those in which the viral proteins are biosynthesised mostly on the endoplasmic reticulum in the cytoplasm.

Table 4. Influence of Potassium Tartrate on the Titre of Viruses *in Vivo*.

Viruses	*Titre reduction factor*								
	% (w/v) potassium tartrate in medium								
	5	*4*	*3*	*2*	*1*	*0.5*	*0.1*	*0.05*	*0.01*
Coxsackie virus	T	T	T	10	1	1	1	1	1
Poliovirus	T	T	T	10	1	1	1	1	1
Herpes virus	T	T	T	1	1	1	1	1	1
Measles virus	T	T	T	1	1	1	1	1	1
Semliki Forest virus	T	T	T	1	1	1	1	1	1

T = toxic for Vero cells

Table 5. Influence of CsCl on the Titre of Viruses *in Vivo*.

Viruses	*Titre reduction factor*								
	% (w/v) CsCl in medium								
	5	*4*	*3*	*2*	*1*	*0.5*	*0.1*	*0.05*	*0.01*
Coxsackie virus	T	T/2	10^2	10	1	1	1	1	1
Poliovirus	T	T/2	10^6	10^5	10^4	10	1	1	1
Herpes virus	T	T/2	10^4	10^4	10^3	10^2	10^2	10	1
Measles virus	T	T/2	10^4	10^4	10^2	10	1	1	1
Semliki Forest virus	T	T/2	10^6	10^5	10^5	10^5	1	1	1

T = toxic for cells, cells disrupted
T/2 = toxic effect on cells, virus growth

3.2.3. *Influence of Potassium Tartrate on the Titre of Viruses in Vivo*

Potassium tartrate is toxic for Vero cells in concentrations above 2% (w/v) as shown in *Table 4*. At concentrations lower than 1% (w/v) there is no effect on the virus titre.

3.2.4. *Influence of CsCl on the Titre of Viruses in Vivo*

Table 5 shows the influence of CsCl on virus titre. CsCl is toxic to cells at concentrations above 3% (w/v). At 1% CsCl the titre of poliovirus is drastically reduced (0.01% recovery) while Coxsackie virus is not affected at all. Caesium ions bind to viral RNA and interfere with viral RNA synthesis and virus maturation (42). The concentration of CsCl is very critical for Semliki Forest virus infectivity while herpes virus is more gradually affected by increasing CsCl concentrations.

3.2.5. *Influence of $SrCl_2$ on the Titre of Viruses in Vivo*

$SrCl_2$ is less toxic than CsCl for tissue culture cells. As shown in *Table 6* the titre of measles virus, a labile enveloped virus, is practically unaffected by up to 4% $SrCl_2$.

3.2.6. *Influence of Nycodenz on the Titre of Viruses in Vivo*

Nycodenz is not toxic for Vero cells in concentrations up to 5%. This is com-

Table 6. Influence of $SrCl_2$ on the Titre of Viruses *in Vivo.*

Viruses	*Titre reduction factor*								
	% (w/v) $SrCl_2$ in medium								
	5	4	3	2	1	0.5	0.1	0.05	0.01
Coxsackie virus	10^4	10^2	10^2	1	1	1	1	1	1
Poliovirus	10	10	10	1	1	1	1	1	1
Herpes virus	10^4	10^3	1	1	1	1	1	1	1
Measles virus	10	1	1	1	1	1	1	1	1
Semliki Forest virus	10^3	10	10	1	1	1	1	1	1

Table 7. Influence of Nycodenz on the Titre of Viruses *in Vivo.*

Viruses	*Titre reduction factor*								
	% (w/v) Nycodenz in medium								
	5	*4*	*3*	*2*	*1*	*0.5*	*0.1*	*0.05*	*0.01*
Coxsackie virus	1	1	1	1	1	1	1	1	1
Poliovirus	1	1	1	1	1	1	1	1	1
Herpes virus	10^2	10^2	10^2	10^2	10^2	10	1	1	1
Measles virus	1	1	1	1	1	1	1	1	1
Semliki Forest virus	1	1	1	1	1	1	1	1	1

parable to sucrose and metrizamide. *Table 7* shows that only the titre of herpes virus is affected by Nycodenz; the mechanism of action is not known. The titration of animal RNA viruses in tissue culture medium containing different gradient media up to 5% (w/v) shows that density gradient fractions can easily by titrated after a ten-fold dilution and thus it is unnecessary to remove the gradient medium by dialysis if the original virus titre exceeds $10^4 TCD_{50}$/ml except for poliovirus, herpesvirus, measles virus and Semliki Forest virus in CsCl. Nycodenz, sucrose and metrizamide have the least effect on the virus titre.

4. EQUILIBRIUM CENTRIFUGATION OF VIRUSES IN DIFFERENT DENSITY-GRADIENT MEDIA

After centrifugation all titrations of density-gradient fractions were done directly without dialysis according to the method described in Section 3.

4.1. Poliovirus

Figure 4 shows the equilibrium centrifugation banding pattern of poliovirus in different media as determined by infectivity. In CsCl, $SrCl_2$ and Nycodenz gradients a single peak of infectivity is obtained with densities of 1.34 g/ml, 1.28 g/ml and 1.31 g/ml, respectively. These results, summarised in *Table 8*, are interesting for taxonomic purposes. The density in CsCl agrees with the accepted value of the density for the genus enterovirus (1).

In potassium tartrate, metrizamide and sucrose gradients, more than one peak of infectivity is found. The density of the main peak of infectivity in

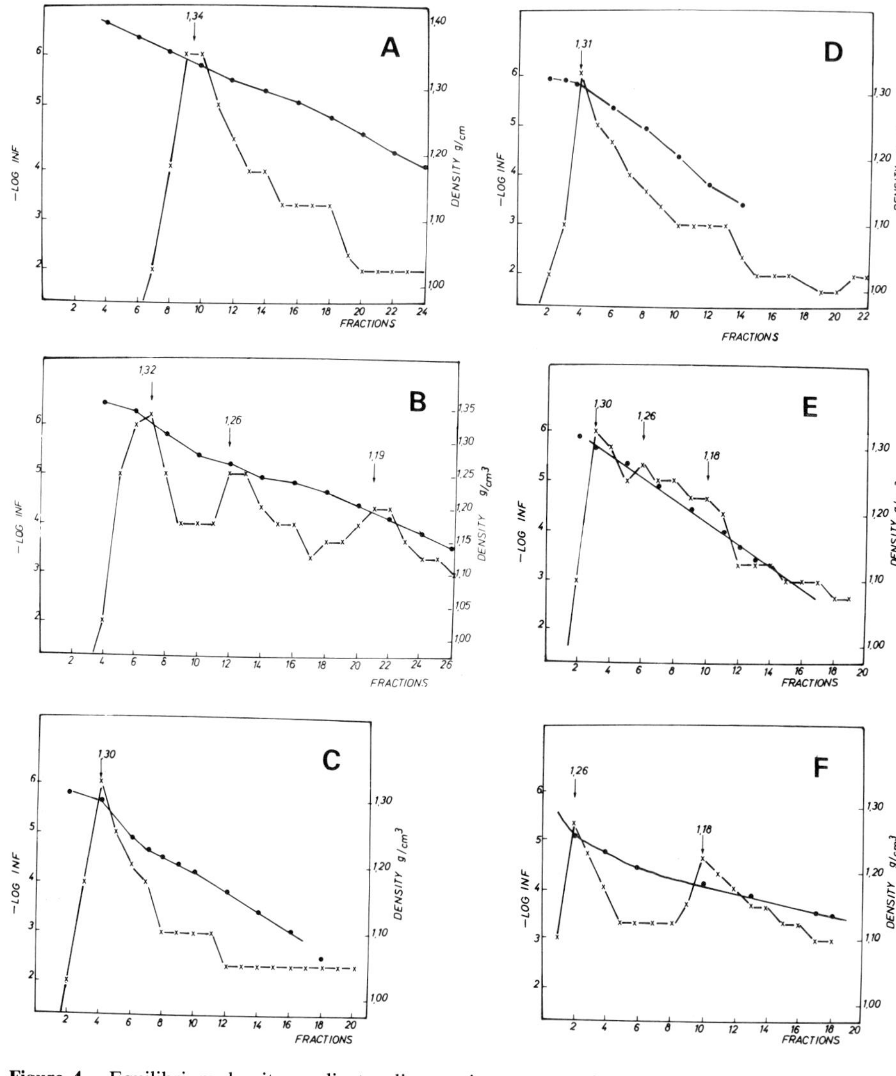

Figure 4. Equilibrium density-gradient sedimentation pattern of poliovirus in different media. Poliovirus was centrifuged for 18 h at 200 000xg at 4°C. The log of virus infectivity (×——×) and density (●——●) of each fraction was measured in A) CsCl (RSB); B) metrizamide/D_2O (RSB); C) Nycodenz (TNE); D) Nycodenz (RSB); E) metrizamide/H_2O (TNE); F) sucrose (TNE).

metrizamide gradients is the same as in Nycodenz gradients. However in metrizamide gradients, in both H_2O and D_2O, poliovirus forms broad diffuse bands. In sucrose, poliovirus bands at lower densities, forming the two peaks at 1.18 g/ml and 1.26 g/ml, as compared to the banding densities in metrizamide and Nycodenz gradients. In order to elucidate the reason for the different infectivity peaks in certain media, labelled poliovirus was purified by rate-zonal sucrose gradient (15%-30%) centrifugation as shown in *Figure 5*. The poliovirus peak (150S) fractions were pooled and centrifuged at 50 000xg

Table 8. The Buoyant Densities of Different RNA Viruses as Determined by Infectivity Measurements.

Medium	*Polio-virus*	*Coxsackie virus*	*Semliki Forest virus*	*Measles virus*
CsCl (RSB)	1.34	1.33	1.26-1.11	1.21
Potassium tartrate (RSB)	1.34-**1.26**	1.26	1.20	1.20
$SrCl_2$ (RSB)	1.28	**1.28**-1.25-1.17	1.26-1.13	(1.12)
$SrCl_2$ (TNE)	1.28	**1.29**-1.25-1.17		
Metrizamide (RSB)	**1.31**-1.18	1.17	1.20-1.12	(1.19)
Metrizamide (TNE)	**1.31**-1.26	1.19	1.20	(1.19)
Metrizamide/D_2O (RSB)	**1.31**-1.26-1.19	1.19	1.20-**1.17**	1.18
Nycodenz (RSB)	1.31	1.26-**1.18**		1.18
Nycodenz (TNE)	1.30	**1.26-1.18**	1.18	1.18
Sucrose (RSB)	1.26-**1.18**	1.26-**1.18**	1.18	1.18
Sucrose (TNE)	**1.26**-1.18	**1.25-1.18**	1.18	1.18

Densities in bold figures indicate the major peak (90% infectivity recovery)
( ) = depends on the strain used
Densities expressed in g/ml.

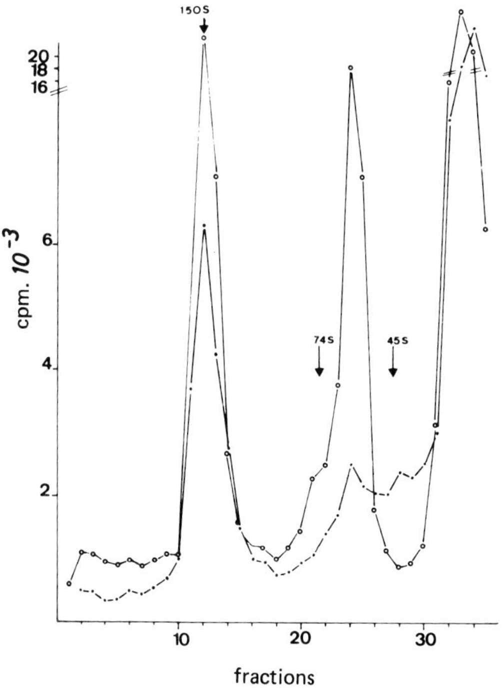

Figure 5. Purification of poliovirus by rate-zonal sedimentation centrifugation in sucrose gradients. Poliovirus in infected Vero cell extracts and labelled with ^{14}C-valine and ^{3}H-uridine were purified by centrifugation in 15-30% sucrose gradients. The arrows indicate the 150S poliovirus peak and ribosomal subunits. The distributions of ^{14}C-valine ○——○: cpm x 10^{-3} and ^{3}H-uridine ○——○: cpm x 10^{3} was determined by liquid scintillation counting.

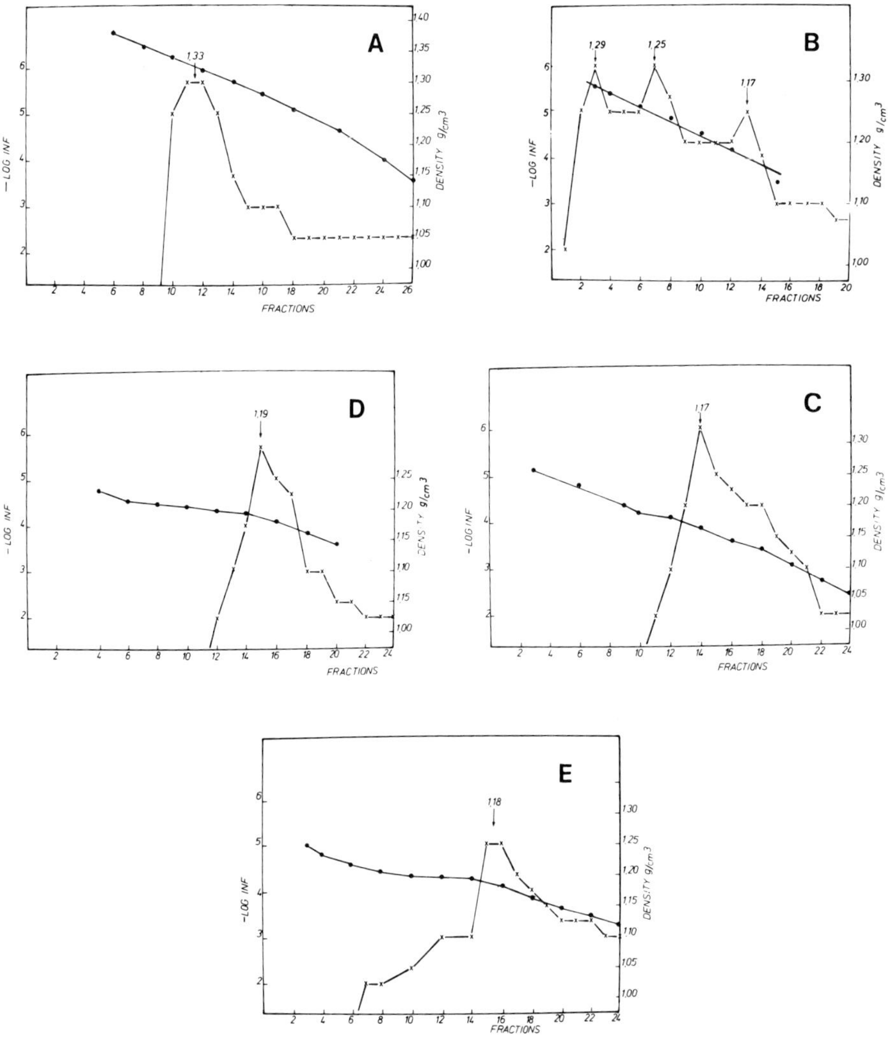

Figure 6. Equilibrium density sedimentation pattern of Coxsackie virus in different media. Coxsackie virus was centrifuged for 18 h at 200 000xg at 4°C. The log of the infectivity (●——●) and density (×——×) of each fraction was measured after centrifugation in A) CsCl (RSB); B) $SrCl_2$ (TNE); C) metrizamide (RSB); D) metrizamide/D_2O (RSB); E) metrizamide (TNE).

for 3 h at 4°C. The pellet was resuspended in 1 ml of RSB buffer and equilibrium centrifugation was done in metrizamide (RSB) and sucrose (TNE) gradients. In both cases only one peak of infectivity was detected at a density of 1.31 g/ml in metrizamide and 1.26 g/ml in sucrose gradients. Evidently the minor peaks originally found at lower densities represent complexes of poliovirus and cellular plasma membranes which sediment quickly (>1000S) and are not disrupted in sucrose and metrizamide media (39). Only a minor part of infectious poliovirus (<10%) is bound to cellular material in Vero cell extracts.

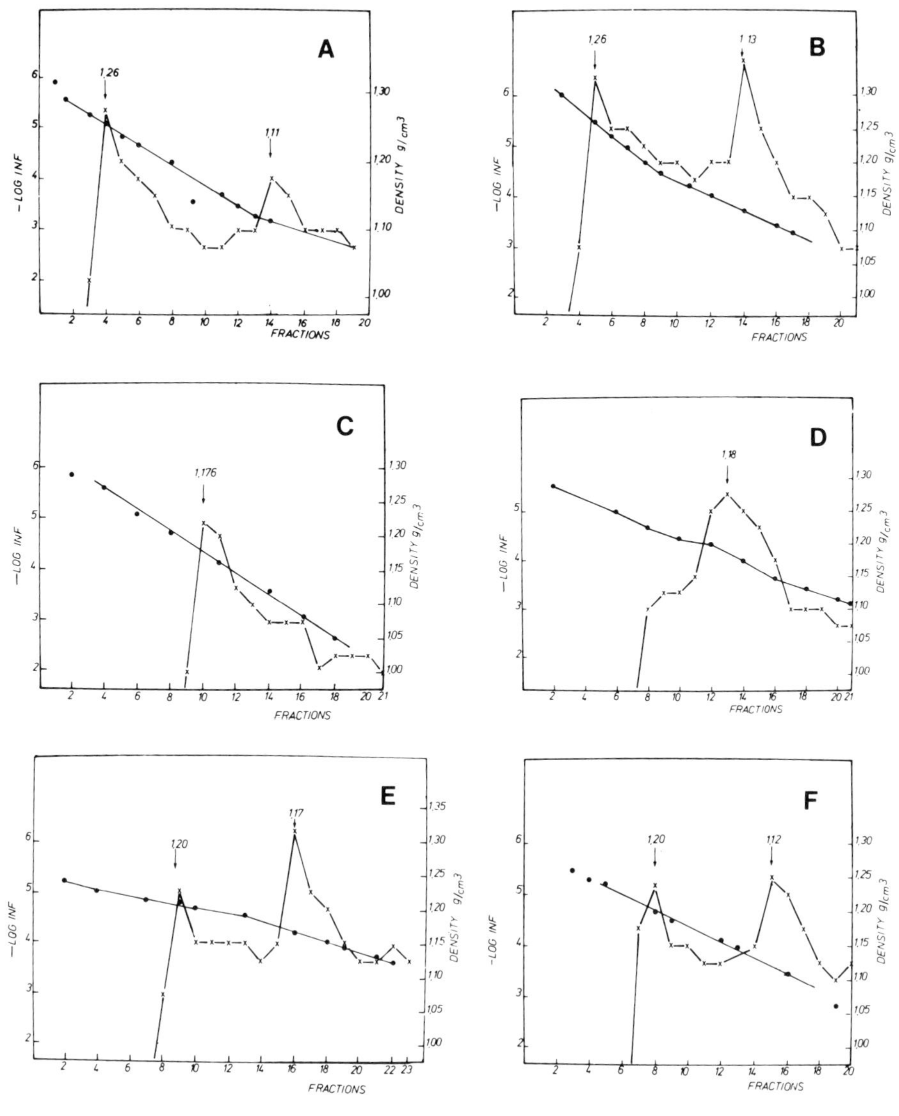

Figure 7. Equilibrium density sedimentation pattern of Semliki Forest virus in different media. Semliki Forest virus was centrifuged for 18 h at 200 000xg at 4°C. The log of the virus infectivity (×——×) and density (○——○) of each fraction were measured after centrifugation in A) CsCl (RSB); B) $SrCl_2$ (RSB); C) Nycodenz (TNE); D) sucrose (RSB); E) metrizamide/D_2O (RSB); F) metrizamide (RSB).

4.2. Coxsackie Virus

Figure 6 shows the density-gradient pattern of Coxsackie virus and *Table 8* summarises the banding densities of this virus in the different media. In CsCl, potassium tartrate and metrizamide (H_2O and D_2O) Coxsackie virus bands as a single peak at densities of 1.33 g/ml, 1.26 g/ml, 1.17 g/ml and 1.19 g/ml, respectively. In $SrCl_2$ three different peaks can be distinguished while in

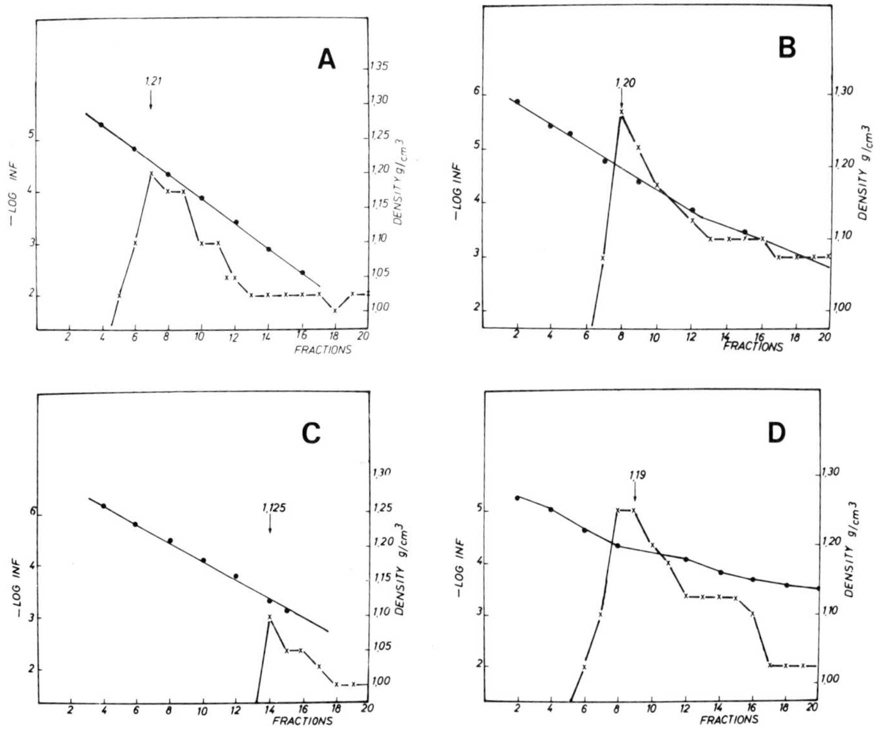

Figure 8. Equilibrium density sedimentation pattern of measles virus. Measles virus Edmonston A and B were centrifuged in different media for 18 h at 200 000xg at 4°C. The density (○——○) and log of virus infectivity (×——×) of each fraction were measured after centrifugation of A) Edmonston A in CsCl (RSB); B) Edmonston A in potassium tartrate (RSB); C) Edmonston A in metrizamide (RSB); D) Edmonston B in metrizamide (RSB).

Nycodenz and sucrose gradients two peaks with similar densities are observed. The density of Coxsackie virus is similar to that of poliovirus in CsCl and potassium tartrate, but the two differ in the other media. To investigate this phenomenon further Coxsackie virus peaks from a metrizamide gradient were analysed in a CsCl gradient. In this case, only a single peak with density of 1.33 g/ml is observed. On the other hand, a peak of Coxsackie virus from a CsCl gradient analysed in a metrizamide gradient in H_2O results in a single peak with a density of 1.31 g/ml which corresponds to the density of poliovirus in metrizamide. These observations suggest that infectious Coxsackie virus is mainly bound to the endoplasmic reticulum of Vero cells. These complexes remain intact in metrizamide, Nycodenz and $SrCl_2$ gradients. Hence it appears that the actual buoyant densities of pure infectious poliovirus and Coxsackie virus are identical, but that some media such as metrizamide, Nycodenz and $SrCl_2$ do not disrupt the binding of the virus particles to the cellular material.

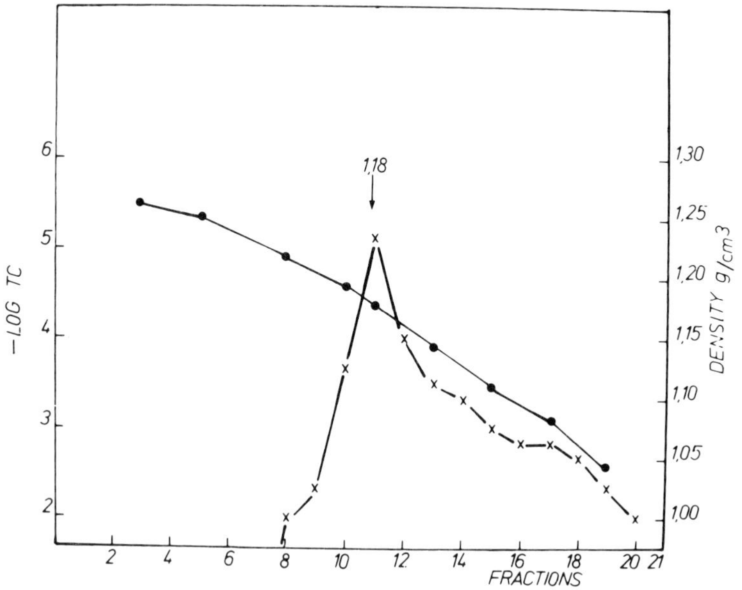

Figure 9. Centrifugation of measles virus Edmonston A in metrizamide/D_2O (RSB). Measles virus Edmonston A was centrifuged in metrizamide/D_2O gradients in RSB buffer at 200 000xg for 18 h at 4°C. The log of the infectivity (○——○) and density (×——×) of each fraction were measured.

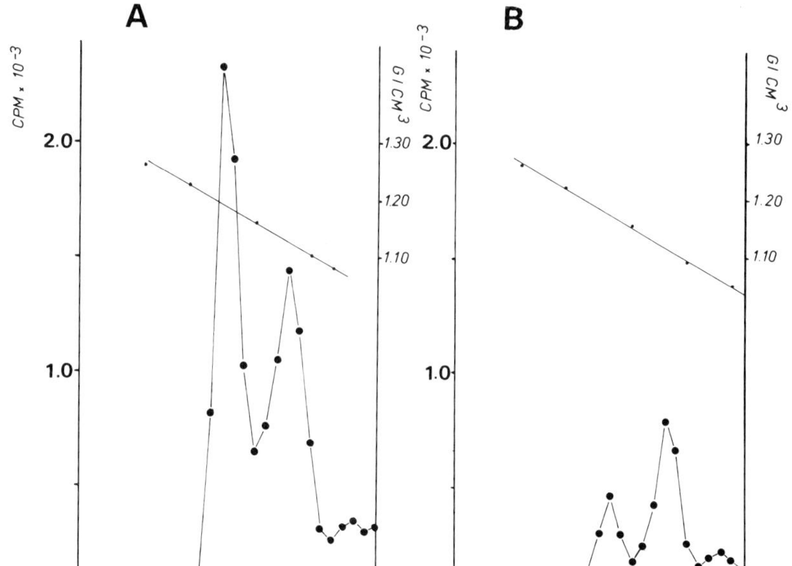

Figure 10. Equilibrium density-gradient centrifugation of ^{35}S-methionine labelled measles virus. Samples of (A) infected Vero cell extract and (B) non-infected Vero cell extract were centrifuged in sucrose gradients in RSB buffer. The radioactivity (○——○) cpm x 10^{-3} and density (———) of each fraction were measured.

4.3. Semliki Forest Virus

A single peak of infectivity is obtained in potassium tartrate, Nycodenz and sucrose gradients *(Figure 7)*. Semliki Forest virus, purified by rate-zonal sedimentation in sucrose gradients ($S_{20,w}$ = 280S), yields single bands in sucrose (1.18 g/ml), $SrCl_2$ (1.26 g/ml) and metrizamide (1.20 g/ml) which correspond with earlier observations made using sucrose gradients (13,14,15).

4.4. Measles Virus

Some strains of measles virus are inactivated by metrizamide or $SrCl_2$. The residual infectious Edmonston A virus bands at a lower than usual density of 1.12 g/ml in metrizamide *(Figure 8C)* while Edmonston B virus bands at the normal buoyant density of 1.18 g/ml *(Figure 8D)*. Using metrizamide/D_2O gradients all measles virus strains tested band at a density of 1.18 g/ml *(Figure 9)*; there is less inactivation of the virus because lower metrizamide concentrations are needed to obtain this density. *Figure 10* shows the density-gradient pattern of labelled measles virus *(Figure 10A)* and an identical labelled cell control (non-infected cell extracts) *(Figure 10B)*. In both cases, peaks of labelled material banding at 1.18 g/ml are clearly present. Evidence from marker enzyme activities indicate that endoplasmic reticulum and plasma membranes are associated with the peak fractions banding at 1.18 g/ml (40) which explains the presence of such peaks on density gradients of different viruses when the medium does not disrupt the associations between the virus and the cell membranes. Hence, the apparent buoyant density of semi-purified virus in relatively inert media such as sucrose, metrizamide or Nycodenz must be evaluated carefully.

5. ACKNOWLEDGEMENTS

The author thanks Prof. G. Hatfield (University of Michigan, Ann Arbor, Michigan, U.S.A.) for critical appraisal of the manuscript.

6. REFERENCES

1. Matthews,R.E.F. (1979) *Intervirology*, **12**, 132.
2. Jamison,R.M. and Mayor,H.D. (1966) *J. Bacteriol.*, **91**, 1971.
3. Schaffer,F.L. and Frommhagen,L.H. (1965) *Virology*, **25**, 662.
4. Mattern,C.F.T. (1962) *Virology*, **17**, 520.
5. Frommhagen,L.H. and Martins,M.J. (1961) *Virology*, **15**, 30.
6. Vanden Berghe,D.A., Poliovirus, doctoral thesis, 1972, University of Ghent, Belgium
7. Summers,D.F., Maizel,J.V.,Jr. and Darnell,J.E.,Jr. (1965) *Proc. Natl. Acad. Sci. (USA)*, **54**, 505.
8. Fernandez-Tomas,C. and Baltimore,D. (1973) *J. Virol.*, **12**, 1122.
9. Jacobson,M.F. and Baltimore,D. (1968) *J. Mol. Biol.*, **33**, 369.
10. Delgadillo,R., Influence of glucosamine on virus multiplication, Doctoral thesis, 1981, University of Antwerp, U.I.A., Belgium.
11. Ghendon,Y., Yakobson,E., Mikhejeva,A. (1972) *J. Virol.*, **10**, 261.
12. Vanden Berghe,D.A. (1973) *Biochem. Biophys. Res. Commun.*, **50**, 957.
13. Hilferhaus,J., Köhler,R. and Grushckau,H. (1976) *J. Biol. Standardization*, **4**, 285.
14. Fox,S.M., Birnie,G.D., Martin,E.M. and Sonnabend,J.A. (1968) *J. Gen. Virology*, **2**, 455.
15. Schenk,T.E. and Stoller,V. (1973) *Virology*, **53**, 162.
16. Vanden Berghe,D.A. and Huybreghts,G. (1979) Humoral immunity in Neurological Diseases,

p. 393, Plenum Press, New York.
17. Huybreghts,G., Vanden Berghe,D.A. and Pattyn,S.R. (1980) *Biochem. Biophys. Res. Commun.,* **93**, 999.
18. Rickwood,D. (1976) *Biological Separations in Iodinated Density Gradient Media.* IRL Press Ltd., Oxford & Washington.
tion Retrieval Ltd., London and Washington.
19. Cash,P., Preston,C.M. and Pringle,C.R. (1979) *Virology,* **96**, 442.
20. Wunner,W.H. and Pringle,C.R. (1976) *Virology,* **73**, 228.
21. Robin,J. and Rodrigue,A. (1978) *Can. J. Microbiol.,* **24**, 1335.
22. Trudel,M., Marchessault,F. and Payment,P. (1981) *J. Virol. Meth.,* **2**, 141.
23. Takahashi,T., Kaga,K., Akakane,Y., Yamashita,T., Miyakawa,Y. and Mayumi,M. (1980) *J. Med. Microbiol.,* **13**, 163.
24. Al-Shammari,A.J.N. and Smith,P.F. (1981) *J. Gen. Virol.,* **54**, 455.
25. Serwer,Ph. (1980) *J. Mol. Biol.,* **138**, 65.
26. Kongsvik,J.R., Hopkins,M.S. and Ellem,K.A.O. (1979) *Virology,* **96**, 646.
27. Gschwender,H.H. and Traub,P. (1978) *Arch. Virol.,* **56**, 327.
28. Zyliez,M. and Taylor,K. (1981) *Eur. J. Biochem.,* **113**, 303.
29. Hudson,J.B., Flawith,J. and Dimmock,N.J. (1978) *Virology,* **86**, 167.
30. Twigden,S.J. and Horisberger,M.A. (1980) *Experientia,* **36**, 1449.
31. Kedinger,C., Brison,O., Perrin,F. and Wilhelm,J. (1978) *J. Virol.,* **26**, 364.
32. Blanchard,J.M. and Weber,J. (1981) *J. Mol. Biol.,* **7**, 107.
33. Everitt,E., Meador,S.A. and Levine,A.S. (1977) *J. Virol.,* **21**, 199.
34. Hronovsky,V., Plaisner,V. and Benda,R. (1975) *Acta Virol.,* **19**, 150.
35. Vanden Berghe,D.A. and Boeyé,A. (1973) *Arch. für die gesammte Virusforschung,* **40**, 215.
36. Reed,L.J. and Muench,H. (1938) *Am. J. Hyg.,* **27**, 493.
37. Vanden Berghe,D.A., Ieven,M., Mertens,F. and Vlietinck,A. (1978) *Lloydia,* **41**, 463.
38. Vanden Berghe,D.A., Van der Groen,G. and Pattyn,S.R. (1976) *Arch. Internat. Physiol. Biochem.,* **84**, 670.
39. Van der Groen,G., Vanden Berghe,D.A., Delgadillo,R. and Pattyn,S.R. (1976) *Arch. Internat. Physiol. Biochem.,* **84**, 671.
40. Vanden Berghe,D.A., unpublished results.
41. Delgadillo,R.A. and Vanden Berghe,D.A. (1981) Fifth Int. Congress Virol., Strasbourg, France, P1/32, p. 59.
42. Mapoles,J.A., Anderegg,J.W. and Reuckert,R.R. (1978) *Virology,* **90**, 103.

APPENDIX I

Enzymic and Chemical Assays Compatible with Iodinated Density-gradient Media

J. Graham and T.C. Ford

1. ASSAYS OF MARKER ENZYMES

    Ouabain-sensitive Na^+/K^+-ATPase
    5′-Nucleotidase
    Aminopeptidase
    Succinate-cytochrome c reductase
    NADPH-cytochrome c reductase and NADH-cytochrome c reductase
    Phosphatase assays
    Glucose-6-phosphatase
    Galactosyl transferase
    Catalase

2. ASSAYS FOR NUCLEIC ACIDS

    Diphenylamine assay for DNA
    Methyl green assay for DNA
    DAPI fluorescent assay for DNA
    Ethidium bromide assay for DNA and RNA
    Orcinol assay for RNA

3. ASSAYS FOR PROTEINS

    Amido-black filter assay
    Coomassie blue filter assay
    Fluorescamine assay
    Coomassie blue solution assay

4. ASSAY FOR POLYSACCHARIDES AND SUGARS

    Phenol-H_2SO_4 assay.

1. ASSAYS OF MARKER ENZYMES

1.1. Ouabain-sensitive Na^+/K^+-ATPase

Avruch, J. and Wallach, D. F. H. (1971) *Biochim. Biophys. Acta,* **233**, 334-347

Solutions

10 mM [γ-^{32}P]-ATP, 1 μCi/ml (37kBq/ml);
1.5 M NaCl;
10 mM $MgCl_2$;
0.3 M KCl;
200 mM Tris-HCl(pH 7.4);
10 mM ouabain;
0.01 M HCl, 1 mM sodium phosphate;
activated charcoal suspension (Norit): contains 4%v/v Norit A, 0.1 M HCl, 0.2 mg/ml bovine serum albumin, 1 mM sodium phosphate, 1 mM sodium pyrophosphate.

Assay method

Set up assay tubes containing 0.1 ml of each of the ATP, NaCl, $MgCl_2$, KCl and Tris-HCl (pH 7.4) solutions and 0.45 ml H_2O. Add 50 μl of sample suspension (0.2-1.0 mg/ml protein) to each tube in the presence or absence of 10 μl of ouabain solution. Set up a blank assay as above, but without the sample suspension, to measure the background rate of ATP hydrolysis. Incubate the assay mixture for 60 min at 37°C and stop the reaction by the addition of 3.0 ml of the charcoal suspension. Unhydrolysed ATP is adsorbed onto the activated charcoal. Stand the tubes at 4°C for 30 min with occasional shaking and then filter the solution through a 2.5 cm Whatman glass-fibre filter disc (GF/C), in a Millipore micro-analysis filter-holder, directly into a scintillation vial. Wash the residue twice with 3 ml of 0.01 M HCl containing 1 mM phosphate. Measure the non-adsorbed ^{32}P-phosphate by Cerenkov counting in a liquid-scintillation counter.

Notes

1.2. 5′-Nucleotidase

Avruch, J. and Wallach, D. F. H. (1971) *Biochim. Biophys. Acta,* **233**, 334 – 347.

Solutions

10 mM [U-^{14}C]-AMP, 0.5 μCi/ml (18.5 kBq/ml);
1.8 mM $MgCl_2$;
500 mM Tris-HCl (pH 8.0);
0.3 N $ZnSO_4$;
0.3 N $Ba(OH)_2$.
The latter two solutions are best obtained as solutions through Sigma Chemical Co. (Poole, Dorset) as preparation of the solutions from solids has been found to be generally unsatisfactory.

Assay method

Set up assay tubes containing 0.1 ml of each of the AMP, $MgCl_2$ and Tris-HCl (pH 8.0) solutions. Add 50 μl H_2O and 50 μl of sample suspension (0.1-1.0mg/ml protein). Prepare a blank as above, but without the sample suspension, to measure the background rate of AMP hydrolysis.

Incubate the reaction mixtures at 37°C for 60 min and stop the reaction by addition of 0.3 ml of each of the $ZnSO_4$ and $Ba(OH)_2$ solutions (unhydrolysed AMP is precipitated by $BaSO_4$). Stand the tubes at 4°C for 30 min with occasional shaking and centrifuge at 700xg for 20 min. Transfer 0.5 ml of the supernatant to a scintillation vial, add a suitable water-soluble scintillator (e.g. Beckman Ready-Solv. HP) and measure the [U-^{14}C]adenosine using a liquid-scintillation counter.

Notes

1.3. **Aminopeptidase**

Wachsmuth, E. D., Fritze, E. and Pfleiderer, G. (1966) *Biochemistry,* **5**, 169–174.

Solutions

25 mM leucine *p*-nitroanilide;
50 mM sodium phosphate (pH 7.2).

Assay method

Use a 1.0 ml cuvette and add 1.0 ml of phosphate buffer, 0.1 ml leucine *p*-nitroanilide and 0.1 ml of the sample. Monitor the absorbance at 405 nm continuously using a chart-recorder with a full scale deflection equivalent to 0.2 absorbance units. The molar extinction coefficient of *p*-nitrophenol at 405 nm is 9620.

Notes

1.4. **Succinate-cytochrome c Reductase**

Method 1.

Mackler, B., Collip, P. J., Duncan, H. M., Rao, N. A. and Heunnekens, F. M., (1962) *J. Biol. Chem.* **237**, 2968 – 2974.

Solutions

50 mM sodium phosphate buffer (pH 7.4);
100 mM potassium cyanide (*CAUTION*, very poisonous);
660 mM sodium succinate;
cytochrome c (12.5 mg/ml).
Make up all reagents in the phosphate buffer.

Assay method

In a 1 ml cuvette, make up an assay mixture containing, 1.0 ml phosphate buffer, 20 μl potassium cyanide, 50 μl cytochrome c and 10 – 50 μl of the sample. Record the absorbance at 552 nm, using a chart-recorder with a full scale deflection equivalent to 0.2 absorbance units. When the trace is steady, add 0.1 ml of sodium succinate solution and record the increase in absorbance due to the reduction of cytochrome c. The molar extinction coefficient of reduced cytochrome c at 552 nm is 27,000.

Notes

The sodium succinate solution may be stored indefinitely at –20°C. The potassium cyanide and cytochrome c solutions must be freshly prepared. **CAUTION; potassium cyanide is extremely poisonous.**
This method can be used with metrizamide but not with Nycodenz.

Method 2.

Modification of the method of Chambers, J. A. A. and Rickwood, D. (1978) in *Centrifugation, A Practical Approach* (Ed. D. Rickwood) p. 43, IRL Press Ltd., Oxford & Washington.

Solutions

20 mM Tris-HCl (pH 7.5), 0.1 mM EDTA;
0.2 M sodium succinate (pH 7.5);
10 mM 2-*p*-iodophenyl-3-*p*-nitrophenyl tetrazolium chloride (INT).

Assay method

Incubate the following incubation mixture in Eppendorf microcentrifuge tubes: 0.5 ml Tris-HCl, EDTA, 0.1 ml sodium succinate, 0.05 ml INT and 0.1 ml of the sample (added last to start the reaction).
Incubate the mixture for 5 min at room temperature (20 – 23°C) and stop the

reaction by the addition of 0.025 ml of 2% SDS. Centrifuge the tubes for 2 min at top speed in the microcentrifuge and remove the supernatant. Measure the absorbance of the supernatant at 500 nm.

Notes

1.5 NADPH-cytochrome c Reductase and NADH-cytochrome c Reductase

Williams, C. H. and Kamin, H. (1962) *J. Biol. Chem.* **237**, 587–595.

Solutions

For NADPH-cytochrome c reductase:
NADPH (2 mg/ml);
cytochrome C (25 mg/ml);
10 mM EDTA;
all made up in 50 mM sodium phosphate (pH 7.7).
For NADH-cytochrome c reductase:
NADH (1 mg/ml);
cytochrome c (25 mg/ml);
both made up in 50 mM sodium phosphate (pH 7.2).

Assay method

Use a 1 ml cuvette and add 0.9 ml of the appropriate buffer, 50 μl cytochrome c, 10 μl EDTA (for the NADPH-cytochrome c reductase only) and 10–20 μl of sample. Using a chart-recorder (full scale deflection equivalent to 0.2 absorbance units) record the absorbance at 552 nm until steady, then add 0.1 ml of NADPH or NADH as appropriate and measure the increase in absorbance due to the reduction of cytochrome c. The molar extinction coefficient of reduced cytochrome c is 27 000 at 552 nm.

Notes

NADPH and NADH solutions must be freshly prepared, protected from light and kept at 0°C until used.

1.6 **Phosphatase Assays**

Engstrom, L. (1961) *Biochim. Biophys. Acta,* **52**, 36 – 48.

1.6.1. *Alkaline Phosphatase*

Solutions

16 mM *p*-nitrophenyl phosphate;
50 mM sodium borate buffer (pH 9.8);
1.0 M $MgCl_2$;
0.25 M NaOH.

Assay method

Make up a stock assay mixture of 5.0 ml substrate, 5.0 ml sodium borate buffer and 20 μl $MgCl_2$. Take 0.2 ml of the assay mixture and add 50 μl of sample (0.1 – 1.0 mg/ml protein) and incubate at 37°C for 60 min. To stop the reaction add 0.6 ml of 0.25 M NaOH and centrifuge at 700xg for 20 min. Measure the absorbance of the supernatant at 410 nm.

1.6.2. *Acid Phosphatase*

Carry out the assay as for alkaline phosphatase (1.6.1) substituting 180 mM sodium acetate buffer (pH 5.0) for the borate buffer and omit the $MgCl_2$.

Notes

1.7. **Glucose-6-phosphatase**

Aronson, N. N. and Touster, O. (1974) *Methods in Enzymol.* **31**, 90 – 102.

Solutions

Prepare the assay solution by mixing 0.1 M glucose-6-phosphate (sodium salt), 35 mM histidine buffer (pH 6.5) and 10 mM EDTA in the ratio 2:5:1:1;
8%(w/v) trichloroacetic acid (TCA);
2.5% ammonium molybdate in 5 N H_2SO_4;
reducing solution: 0.5 g 1-amino-2-naphthol-4-sulphonic acid, 1.0 g anhydrous Na_2SO_3, in 200 ml of 15% $NaHSO_3$.

Assay method

Add 50 μl of sample (approx. 1 mg/ml protein) to 0.45 ml of the assay solution and incubate at 37°C for 30 min. For every sample prepare a blank containing no substrate, together with a reagent blank to determine the background hydrolysis of the substrate. Stop the reaction by adding 2.5 ml of cold (8%w/v) TCA and keep the acidified sample at 0°C for 20 min. Centrifuge at 700xg for 20 min, take 1.0 ml of the supernatant and add 1.15 ml distilled water. Add 0.25 ml of the ammonium molybdate solution and 0.1 ml of the $NaHSO_3$ reducing solution. Measure the absorbance after 10 min at 820 nm.

Notes

The reducing solution should be freshly prepared each time.

1.8. **Galactosyl Transferase**

Fleisher, B., Fleisher, S. and Ozawa, H. (1969) *J. Cell Biol.* **43**, 59–79.

Solutions

40 mM Hepes-NaOH (pH 7.0);
300 mM $MnCl_2$;
30 mM mercaptoethanol;
1.0 mg/ml uridine diphospho-[6-^{3}H]-galactose, 2 μCi/ml (74 kBq/ml);
400 mM N-acetylglucosamine;
300 mM EDTA (pH 7.4).

Assay method

Set up assay tubes containing 20 μl Hepes-NaOH buffer, 10 μl of each of the $MnCl_2$, mercaptoethanol and uridine diphosphogalactose solutions and 10 μl of sample (20 – 50 μg protein), in the presence or absence of 10 μl of 400 mM N-acetylglucosamine as an acceptor. At the same time, set up blanks with no sample added. Incubate the mixtures at 37°C for 60 min and stop the reaction by the addition of 20 μl of 300 mM EDTA (pH 7.4). Cool the tubes to 4°C. Prepare 2 cm Dowex-1 (chloride form) columns in Pasteur pipettes and wash them through with three volumes of water. Apply each incubation mixture to separate Dowex-1 columns and wash each twice with 0.5 ml water. Collect the eluant in scintillation vials, add a suitable water-soluble scintillation fluid (e.g. Beckman Ready Solv. HP) and measure the radioactivity in each case.

Notes

1.9. **Catalase**

Cohen, G., Dembiec, D. and Marcus, J. (1970) *Anal. Biochem.*, **34**, 30 – 38.

Solutions

0.01 M sodium phosphate buffer (pH 7.0);
6 mM H_2O_2 in sodium phosphate buffer;
6 N H_2SO_4;
0.01 N $KMnO_4$.

Assay method

Set up the following tubes at 4°C:
(1) Test; 50 μl sample, 0.5 ml H_2O_2, vortex and after 3 min add 0.1 ml H_2SO_4 and vortex.
(2) Sample blank; 50 μl sample, 0.1 ml H_2SO_4, 0.5 ml H_2O_2, in that order.
(3) Standard; 0.55 ml of buffer, 0.1 ml H_2SO_4.
(4) Reagent blank; 0.1 ml H_2SO_4, 0.5 ml H_2O_2, 0.5 ml buffer.
(5) Spectrophotometer blank; 0.7 ml H_2O, 0.1 ml H_2SO_4, 0.55 ml buffer.
To tubes 1 – 4 add 0.7 ml $KMnO_4$ and read the optical density at 480 nm against tube 5 within 1 min.

Notes

2. CHEMICAL ASSAYS FOR NUCLEIC ACIDS

2.1. **Diphenylamine Assay for DNA**

Schneider, W. C. (1957) *Methods in Enzymol.* **3**, 680 – 684.

Solutions

Diphenylamine reagent: dissolve 1.0 g diphenylamine in 100 ml glacial acetic acid and add 2.75 ml of concentrated H_2SO_4;
DNA standard solution, 500 μg DNA/ml (A_{260} = 11);
20% (w/v) trichloroacetic acid (TCA).

Assay method

Dilute the DNA stock solution to give sample volumes of 1.0 ml in 5% TCA containing 0 – 200 μg DNA. In the case of metrizamide gradients remove the metrizamide by precipitating the DNA; to do this add an equal volume of cold 20% (w/v) TCA. Pellet the DNA by centrifugation at 1000xg for 20 min and wash the pellet twice with cold 10% TCA to remove all of the metrizamide, which interferes with this assay. Hot-acid digest the standards and gradient fractions in 5% TCA in a water bath at 90°C for 20 min to solubilise the DNA. To 1.0 ml samples add 2.0 ml of the diphenylamine reagent and incubate in a water bath at 100°C for 10 min. Cool to room temperature and measure the optical density at 595 nm using 2.0 ml of diphenylamine reagent and 1.0 ml 5% TCA as the blank.

Notes

This assay cannot be used for metrizoate gradient fractions. Do not store the diphenylamine reagent in the cold since it solidifies.

2.2. **The Methyl Green Assay for DNA**

Peters, D. L. and Dahmus, M. E. (1979) *Anal. Biochem.* **93**, 306 – 311.

Solutions

Methyl green reagent: dissolve methyl green (Kodak-Eastman, Rochester, NY, USA) in 100 mM Tris-HCl (pH 7.9) to a concentration of 0.01% (w/v). To remove the contaminating crystal violet present in the solution, extract the solution with an equal volume of water-saturated chloroform. Allow the mixture to separate in a separating funnel and draw off the chloroform in the lower part of the funnel together with the dissolved crystal violet. Repeat if necessary, using fresh chloroform, until the chloroform layer is clear.
DNA standard solution (500 μg/ml);
Proteinase K solution, 1 mg/ml (predigested at 37°C for 30 min).

Assay method

Dilute the DNA stock solution to give 0.2 ml volumes containing 0 – 100 μg DNA, or make up gradient fractions to 0.2 ml. Add 5 μg (5 μl) of proteinase K to each fraction and incubate at about room temperature for 60 min. Add 1.0 ml of the methyl green reagent to each fraction and incubate either overnight at room temperature (20°C) or for 2 – 3 h at 40 – 45°C. The optical density of the solutions is measured at 640 nm. Prepare a blank using 0.2 ml of distilled water (treated in the same way as the sample fractions) and 1.0 ml of the methyl green reagent.

Notes

In the absence of DNA, the blue colour of the reagent fades, but in the presence of DNA, the colour is retained in proportion to the amount of DNA present. The assay is compatible with samples containing Nycodenz, metrizamide or metrizoate. Methyl green from some sources may not give consistent results.

2.3. **DAPI Fluorescent Assay for DNA**

Brunk, C. F., Jones, K. C. and James, T. W. (1979) *Anal. Biochem.* **92**, 497 – 500.

Solutions

DAPI reagent: 100 mM NaCl, 10 mM EDTA, 10 mM Tris-HCl (pH 7.0) containing 100 ng/ml 4′,6-diamidino-2-phenylindole (DAPI);
DNA standard solution (20 μg/ml).

Assay method

The fluorescence of 3.0 ml of the DAPI solution is determined in the absence of DNA (excitation wavelength 360 nm and emission wavelength 450 nm). Add 15 μl of the DNA standard solution and remeasure the fluorescence; add a further 15 μl DNA solution and again measure the fluorescence, repeat twice more to obtain four points for a standard curve. Take four further fluorescence measurements after each addition of four 15 μl aliquots of the DNA sample solution. Calculate the DNA content of the sample material by comparing the fluorescence yield of the standards and sample material.

Notes

When DAPI is complexed with DNA, the fluorescence of the complex is enhanced about 20-fold as compared to the fluorescence of the dye alone. The fluorescence is not affected at pH values between pH 5 and pH 10. Cations, especially divalent or heavy metal ions, cause significant quenching of the fluorescence and quenching is also observed at low ionic strengths (hence the composition of the buffer used). Quenching is found to increase with decreasing temperature and therefore it is necessary that all measurements are made at a uniform temperature. The binding of DAPI is highly specific for A – T base pairs hence it is essential that the DNA of the standard and sample have the same base composition.

2.4. **Ethidium Bromide Assay for DNA and RNA**

Karsten, U. and Wollenberger, A. (1972) *Anal. Biochem.* **46**, 135 – 148.
Karsten, U. and Wollenberger, A. (1977) *Anal. Biochem.* **77**, 464 – 469.

Solutions

Phosphate buffered saline (PBS): contains 0.1 g $CaCl_2$, 0.2 g KCl, 0.2 g KH_2PO_4, 0.1 g $MgCl_2.6H_2O$, 8.0 g NaCl, 1.15 g Na_2HPO_4, adjusted to pH 7.5 and made up to 1 litre;
ethidium bromide (25 μg/ml) in PBS;
heparin (25 μg/ml) in PBS;
RNase (50 μg/ml) in PBS (heat to 100°C for 10 min to destroy any DNase activity);
DNA standard solution (25 μg/ml DNA) in PBS. Stock DNA solutions should be stored frozen. Before use dilute the solution by addition of four volumes of PBS.
The sample material should be either in PBS or another appropriate buffer.

Assay method

Make up the following assay mixtures in 3 ml cuvettes:
(1) Standard: 0.5 ml DNA standard solution, 0.5 ml heparin solution, 1.0 ml PBS.
(2) Blank I: 0.5 ml heparin solution, 1.5 ml PBS.
(3) Blank II: 2.5 ml PBS.
(4) Sample (DNA + RNA): 0.5 ml homogenate, 0.5 ml heparin solution, 1.0 ml PBS.
(5) Sample (DNA only): 0.5 ml homogenate, 0.5 ml heparin solution, 0.5 ml RNase solution, 0.5 ml PBS.
(6) Sample (background correction): 0.5 ml homogenate, 2.0 ml PBS.

Incubate mixtures (1 – 6) in a waterbath at 37°C for 20 min then add 0.5 ml of ethidium bromide solution to tubes (1), (2), (4) and (5). Full intensity of fluorescence is attained within 60 sec of the addition of ethidium bromide and remains constant for at least 60 min. Briefly stir the reaction mixture before taking any measurements. The fluorescence is measured using an excitation wavelength of 360 nm and an emission wavelength of 580 nm. The fluorescence of the standard is set at 100. Temperature during the measurements is critical and should be kept constant to within ±0.5°C.

Calculation of nucleic acid content

DNA

$$A_{dna} = \frac{A_{std} (F_{(5)} - F_{(2)} - F_{(6)} + F_{(3)})}{F_{(1)} - F_{(2)}}$$

A_{dna} = amount of DNA/mixture (μg). A_{std} = amount of standard DNA/mixture. F = fluorescence intensity (units).

RNA

$$A_{rna} = \frac{A_{std}\,(F_{(4)} - F_{5})}{0.46\,(F_{(1)} - F_{(2)})}$$

The factor 0.46 is empirically derived from the ratio of the fluorescence yield of RNA and DNA with ethidium bromide. Alternatively, an RNA standard solution can be used.

Notes

Up to 5 μg/ml DNA or RNA fluorescence increases linearly with concentration. Check the RNase solution for fluorescence and subtract from the reading mixture (5).

2.5. Orcinol Assay for RNA

Schneider, W. C. (1957) *Methods in Enzymol.* **3**, 680 – 684.

Solutions

Orcinol reagent: 0.5 g orcinol and 0.25 g $FeCl_3.6H_2O$ in 50 ml of concentrated HCl (the reagent should be freshly prepared and kept at 4°C until required); RNA standard solution (250 μg/ml) in distilled water; trichloroacetic acid 20% (w/v) (TCA) solution.

Assay method

Dilute the RNA stock solution to give 0.5 ml volumes containing 0 – 250 μg RNA for the preparation of a standard curve, or dilute gradient fractions to 0.5 ml. Add TCA to a final concentration of 5% (w/v) and incubate each fraction for 20 min at 90°C. Centrifuge for 2 min in a microcentrifuge, remove the supernatant and add to it an equal volume of the orcinol reagent. Incubate for 20 min in a boiling water-bath, cool the solutions and measure the optical densities at 660 nm. Use 1.0 ml of reagent and 1.0 ml of 5% TCA as a blank.

Notes

The yellow colour of the reagent becomes green in the presence of RNA. This assay is incompatible with the presence of metrizoates, also the orcinol reagent reacts with the glucosamide group of the metrizamide molecule. Metrizamide can be removed by acid precipitation (see 2.1.1). Proteins, other sugars and formaldehyde also interfere with the orcinol assay.

3. ASSAYS FOR PROTEINS

3.1. Amido-black Filter Assay

Schaffner, W. and Weissman, C. (1973) *Anal. Biochem.* **56**, 502 – 514.

Solutions

Stain: dissolve 0.1 g Amido-black 10B in methanol/glacial acetic acid/water, 45/10/45 by vol;
destain: methanol/glacial acetic acid/water, 90/2/8 by vol;
eluant solution: (25 mM NaOH and 0.05 mM EDTA in 50% aqueous ethanol;
standard protein solution (150 μg/ml);
stock solution of 60% (w/v) TCA;
1% SDS in 1M Tris-HCl (pH 7.5).

Assay method

Dilute the standard protein solution to a final volume of 0.27 ml containing 0 – 150 μg protein for the standard curve, or dilute gradient fractions to 0.27 ml. Add 0.03 ml of the Tris-SDS solution and then 0.6 ml of 60% TCA to each fraction. Filter each sample through a Millipore filter (0.45 μ pore size) rinse the tube with 0.3 ml 6% TCA and then wash the filter with 2 ml of 6% TCA. Place each filter in stain for 2 – 3 min with gentle agitation and remove to a water rinse for about 30 sec. Pass each filter through three changes of destain, about 1 min per change, rinse again in water and blot with a tissue. Place the filter in a test-tube with 0.6 ml of eluant solution for 10 min, vortexing three times for 2 – 3 sec, add a further 0.9 ml eluant and mix by vortexing. The optical density of the solutions is measured at 630 nm using the eluant solution as a blank.

Notes

Using this assay, no interference is experienced from the presence of gradient solutes unless, like metrizoates, they are not acid soluble; both metrizamide and Nycodenz are acid soluble and thus such gradients may be analysed by this assay. Potassium ions must be excluded from the sample solutions since they precipitate the dodecyl sulphate ions.

3.2. Coomassie Blue Filter Assay

McKnight, G. S. (1977) *Anal. Biochem.* **78**, 86 – 92.

Solutions

Stain: 0.25% (w/v) Coomassie blue, 7.5% acetic acid, 5% methanol in distilled water;
destain: 7.5% acetic acid, 5% methanol;
eluant solution: 0.12 M NaOH, 20% H_2O, 80% methanol:
3.0 M HCl;
20% (w/v) TCA.

Assay method

Cut 25 mm discs from Whatman glass-fibre filters (GF/C). Number the filters along the edge, taking care not to touch the filters with bare fingers. Protein samples are pipetted onto the centre of the discs and allowed to absorb completely (5 – 30 sec). The sample volume to be used depends upon the protein (see notes). Prepare blank filters as described, using distilled water instead of the protein sample. After the absorption period place the filters in 20% TCA at 4°C and swirl gently for several minutes. Transfer the filters individually to the stain solution at 4°C for 20 min. Transfer the filter to destain and swirl for several minutes, then decant the solution and repeat twice with fresh destain. Place the filters on a filtering apparatus and wash several times with destain until the blank filters are completely white. Cut out the stained spots using a 1 cm cork borer and resting the filters on several sheets of filter paper. Put the cut out spots into 1.5 ml microcentrifuge tubes and add 0.6 ml of eluant solution, vortex and stand at room temperature until the colour disappears from the filters (about 5 min). Acidify the eluant with 0.03 ml 3 M HCl (Coomassie blue is colourless in basic solutions), vortex and centrifuge for two minutes in a microcentrifuge to remove any pieces of glass fibre from the solution. Carefully remove the supernatant and measure its absorbance at 590 nm using acidified eluant as a blank.

Notes

This assay is not compatible with metrizoates. The sample volume that can be pipetted onto the filter depends on the salt concentration used and on the particular protein. Some proteins, such as RNase and histones, do not spread and volumes of up to 100 μl can be used to give concentrated spots, but BSA and ovalbumin, for example, diffuse into the filter and in such cases volumes need to be less than 50 μl. In high salt (e.g. 1M NaCl), all proteins bind tightly to the filter and produce a concentrated spot.
The sensitivity of this assay can be increased 3-fold by using 200 μl of eluant solution and 10 μl of 3M HCl.

3.3. **Fluorescamine Assay**

Bohlen, P., Stein, S., Dairman, W. and Udenfriend, S. (1973) *Arch. Biochem. Biophys.* **155**, 213-220.

Solutions

Fluorescamine reagent: 30 mg of fluorescamine in 100 ml dioxane (histological grade);
0.05 M sodium phosphate buffer (pH 8.0).

Assay method

Place the sample, containing 0.5 – 50 μg protein, into a tube and add phosphate buffer to bring the total volume up to 1.5 ml. Add 0.5 ml of the fluorescamine solution with vortexing to ensure vigorous mixing. Rapid mixing is essential for reproducible results since the reagent is hydrolysed rapidly. The pH must be maintained at pH 8 – 9. The reaction takes place at room temperature.
Measure the fluorescence using a fluorimeter with an excitation wavelength of 390 nm and an emission wavelength of 475 nm. More reproducible results are obtained by carrying out direct fluorescence measurements in the reaction tubes, using a cell adaptor, instead of transferring the mixtures to a cuvette. Apparent protein concentrations are calculated by comparison to bovine serum albumin standard solutions. All values should be corrected for a blank, prepared using 1.5 ml of sample buffer.

Notes

Fluorescamine reacts with amino groups (primary amines), thus Tris and other primary amine buffers cannot be used. Large amounts of secondary and tertiary amine buffers also interfere with the assay. Hence inorganic buffers, such as phosphate or borate, are most successful. A modification of this procedure uses gel filtration to remove contaminating low molecular weight amines in the sample which interfere with the assay. Bohlen *et al.* Anal. Biochem. **58**, 559 – 562. Using this modification of the assay as described, it should be noted that contaminants with molecular weights in excess of 300 do interfere due to incomplete separation from the proteins on the Sephadex G-25 columns.

3.4. **Coomassie Blue Solution Assay**

Bradford, M. (1976) *Anal. Biochem.* **72**, 248 – 254.

Solutions

Coomassie blue reagent: dissolve 100 mg Coomassie blue in 50 ml 90% ethanol and add 100 ml 85% (w/v) phosphoric acid, make up to 200 ml with distilled water. This concentrated solution can be stored for at least a month at 5°C. Dilute by the addition of four volumes of distilled water before use. Standard protein solution (100 μg/ml).

Assay method

Dilute the protein standard solution to final volume of 0.1 ml containing 0 – 100μg protein, dilute the sample fractions to 0.1 ml. Add 5.0 ml of the reagent to each fraction and measure the optical density at 595 nm. Measurement should be carried out after 2 min but before 60 min has elapsed after the addition of the reagent. Use 5.0 ml of reagent and 0.1 ml protein buffer as a blank.

Notes

This assay provides a swift and easy determination of protein distribution in a gradient and is compatible with both metrizamide and Nycodenz but not metrizoates.

4. ASSAY FOR POLYSACCHARIDES AND SUGARS

4.1. Phenol-H_2SO_4 Assay

Dubois, M., Gilles, K. A., Hamilton, J. K., Rebers, A. and Smith, F. (1956) *Anal. Chem.* **28**, 350 – 356.

Solutions

80% (w/w) phenol;
concentrated H_2SO_4 (95.5%).
standard sugar solution (100 μg/ml).

Assay method

Use tubes of at least 10 mm diameter to allow good mixing and to minimise heat dissipation. To 1.0 ml of standard or sample sugar solution add 50 μl 80% phenol; add 5.0 ml H_2SO_4 allowing the acid to leave the pipette quickly and fall directly onto the liquid surface, so ensuring good mixing. Let the tubes stand for 10 min, shake and place into a water-bath at 25 – 30°C for 10 – 15 min. Blanks are prepared by substituting water for the sample solution. Measure the optical density at 490 nm in the case of hexoses and at 480 nm in the case of pentoses and uronic acids.

Notes

5% (w/w) phenol may be used in place of 80% phenol, in which case 1.0 ml of phenol solution is added, other volumes as given.

A Bibliography of Key References of the Applications of Nonionic Iodinated Density-gradient Media

1. General

Buoyant density gradient centrifugation in solutions of metrizamide. Hell, A., Rickwood, D. and Birnie, G. D. in *Methodological Developments in Biochemistry,* vol. 4 (Ed. E. Reid) pp. 117-123. Longman, New York (1974).

Metrizamide, a new density gradient medium. Rickwood, D. and Birnie, G.D., FEBS Lett. **50** (1975) 102-110.

Practical aspects of using metrizamide. Birnie, G. D. and Rickwood, D. in *Biological Separations in Iodinated Density-gradient Media* (Ed. D. Rickwood) pp. 193-201. IRL Press Ltd., Oxford and Washington (1976).

Metrizamide interferes with the determination of protein by the Lowry method. Slaby, F. and Farquhar, M. G., Anal. Biochem. **77** (1976) 280-285.

Choice of media for centrifugal separations. Rickwood, D. in *Centrifugation: A Practical Approach* (Ed. D. Rickwood) pp. 15-31. IRL Press Ltd., Oxford and Washington (1978).

A method of DNA quantitation for the localisation of DNA in metrizamide gradients. Peters, D. L. and Dahmus, M. E., Anal. Biochem. **93** (1979) 306-311.

Labelling metrizamide with iodine-125. Sharma, H. L. and Smith, A.G., Acta Radiol. Diagn., **20** (1979) 289-291.

Measurement of protein by dye-binding assay in the presence of metrizamide. Gogstad, O., Anal. Biochem. **106** (1980) 524-528.

Transport phenomena in zonal centrifuge rotors. XIII. Gradient properties of metrizamide. Lu, W. H. and Hsu, H. W., Sep. Sci. Technol. **15** (1980) 1393-1399.

Competitive inhibition of rat-brain hexokinase by 2-deoxyglucose, glucosamine and metrizamide. Bertoni, J. M., J. Neurochem. **37** (1981) 1523-1528.

The isolation of purified neurosecretory vesicles from bovine neurohypophysis using iso-osmolar density gradients. Russell, J. T., Anal. Biochem. **113** (1981) 229-238.

Electrophoresis in density gradients of metrizamide. Serwer, P. and Watson, R.H., Anal. Biochem. **114** (1981) 342-348.

Competitive inhibition of brain hexokinase by metrizamide. Bertoni, J. M. and Steinman, C. G., Neurology (USA) **32** (1982) 320-323.

Formation of isotonic Nycodenz gradients for cell separations. Ford, T. C. and Rickwood, D., Anal. Biochem. **124** (1982) 293-298.

Intracellular hormone receptors; evidence for insulin and lactogen receptors in a unique vesicle sedimenting in lysosome fractions of rat-liver. Khan, M. N., Posner, B. I., Verma, A. K., Khan, R. J. and Bergeron, J. M., Proc. Natl. Acad. Sci. (USA) **78** (1982) 4980-4984.

Biochemical and biophysical characterisation of insulin granules isolated from rat pancreatic islets by an iso-osmotic gradient. Mathews, E. K., McKay, D. B., O'Connor, M. D. L. and Borowitz, J. L., Biochem. Biophys. Acta, **715** (1982) 80-89.

Properties and characteristics of nonionic iodinated gradient media. Rickwood, D., Biol. Cell, **45** (1982) 473.

Characterisation of Nycodenz, a new, nonionic, iodinated density gradient medium. Rickwood, D. and Ford, T. C., J. Cell Biol. **95** (1982) 460a.

Nycodenz; a new nonionic iodinated density gradient medium. Rickwood, D., Ford, T. C. and Graham, J., Anal. Biochem. **123** (1982) 23-31.

Density gradient centrifugation today: metrizamide, Ficoll and Percoll. Sternbach, H., Laborpraxis, **6** (1982) 482-494.

2. Macromolecules

Buoyant densities and hydration of nucleic acids, proteins and nucleoprotein complexes in metrizamide. Birnie, G. D., Rickwood, D. and Hell, A., Biochim. Biophys. Acta, **331** (1973) 293-294.

Isopycnic sedimentation of DNA in metrizamide. Effect of low concentrations of ions on buoyant density and hydration. Birnie, G. D., MacPhail, E. and Rickwood, D., Nucleic Acids Res. **1** (1974) 919-925.

Metrizamide in deuterium oxide, a novel solute for the isopycnic banding of proteins. Hüttermann, A. and Guntermann, U., Hoppe-Seyler's Z. Physiol. Chem. **355** (1974) 1210.

The organisation of proteins in polyoma and cellular chromatin. Louie, A. J., Cold Spring Harbor Symp. Quant. Biol. **39** (1974) 259-266.

Reversible interaction of metrizamide with protein. Rickwood, D., Hell, A., Birnie, G. D. and Gilhuus-Moe, C. C., Biochim. Biophys. Acta, **342** (1974) 367-371.

Induction of glucosidase and synthesis during the cell cycle of Myxobacter AL-1. Guntermann, U., Tan, I. and Hüttermann, A., J. Bacteriol. **124** (1975) 86-91.

Activity, isoenzyme pattern and synthesis of UDP-glucose-4-epimerase during differentiation of *Physarum polycephalum*. Hüttermann, A., Gerbauer, M., Wessel, I. and Hoffmann, W., Biochim. Biophys. Acta, **384** (1975) 493-500.

Metrizamide in deuterium oxide, a novel solute for the separation of labelled and unlabelled proteins by equilibrium density gradient centrifugation. Hüttermann, A. and Guntermann, U., Anal. Biochem. **64** (1975) 360-366.

Interaction of histones with DNA using isopycnic centrifugation in metrizamide gradients. Rickwood, D., Birnie, G. D. and MacGillivray, A. J., Nucleic Acids Res. **2** (1975) 723-733.

Interactions between macromolecules and the gradient forming solute in isopycnic sedimentation. Skerrett, R. J., Biochim. Biophys. Acta, **385** (1975) 28-35.

Isopycnic centrifugation of proteins in metrizamide/deuterium oxide gradients. Hüttermann, A. and Wendlberger, G. in *Biological Separations in Iodinated Density-gradient Media*. (Ed. D. Rickwood) pp. 15-25. IRL Press Ltd., Oxford and Washington (1976).

Studies on metrizamide protein interactions. Hüttermann, A. and Wendlberger-Schieweg, G., Biochim. Biophys. Acta, **453** (1976) 176-184.

Isopycnic centrifugation of nucleic acids and their interaction with nuclear proteins in iodinated density-gradient media. Rickwood, D. in *Biological Separations in Iodinated Density-gradient Media,* (Ed. D. Rickwood) pp. 27-40. IRL Press Ltd., Oxford and Washington (1976).

Application of a metrizamide density gradient to the purification of glycogen particles. Guenard, D., Morange, M. and Buc, H., FEBS Lett. **76** (1977) 262-265.

Quantitative aspects of the binding of nuclear non-histone proteins to DNA as determined by centrifugation in metrizamide gradients. Rickwood, D. and MacGillivray, A. J., Exp. Cell Res. **104** (1977) 287-292.

Separation of macromolecules by isopycnic centrifugation in different media, Chambers, J. A. A. and Rickwood, D. in *Centrifugation: A Practical Approach* (Ed. D. Rickwood) pp. 117-133. IRL Press Ltd., Oxford and Washington (1978).

Density gradient ultracentrifugation method of studying sibiromycin interaction with linear and circular DNA. Kozmion, L. I., Gauze, G. G., Galkin, V. I. and Dudnik, I. V., Antibiotiki, **23** (1978) 771-775.

Complexes of DNA with chromatin proteins investigated by centrifugation in metrizamide. Rzeszowska-Wolny, J., Filipski, J. and Chorazy, M., Studia Biophysica, **67** (1978) 133-134.

Complex of DNA with chromatin protein investigated by isopycnic centrifugation in metrizamide. Rzeszowska-Wolny, J., Filipski, J., Groebner, S. and

Chorazy, M., Nucleic Acids Res. **5** (1978) 4905-4917.

Light mediated activation of nitrate reductase in synchronous chlorella. Tischner, R. and Hüttermann, A., Plant Physiol. **62** (1978) 284-286.

Fractionation of cartilage proteoglycans by metrizamide density-gradient ultracentrifugation. I. On the application of metrizamide. Nakamura, H., Shika Kiso Igakkai Zasshi **21** (1979) 417-427.

Fractionation of cartilage proteoglycans by metrizamide density gradient ultracentrifugation. II. Application for metabolism experiments. Nakamura, H., Shika Kiso Igakkai Zasshi **21** (1979) 428-432.

Isopycnic banding in metrizamide of the uterine cytosol and nuclear estradiol receptor. Baskevitch, P. P. and Rochefort, H., Mol. Cell. Endocrinol. **22** (1981) 195-210.

Isolation of the intercellular glycoproteins of desmosomes. Gorbsky, G. and Steinberg, M., J. Cell Biol. **90** (1981) 243-248.

Separation of mercury substituted RNA synthesized in isolated rat-liver nuclei. Hanausek-Walaszek, M., Walaszek, Z. and Chorazy, M., Mol. Biol. Rep. **7** (1981) 57-62.

A combined theoretical and experimental study of the interaction of metrizamide with proteins. Stimpson, D. I. and Cann, J. R., Arch. Biochem. Biophys. **211** (1981) 403-412.

Fractionation of macromolecules and nucleoproteins. Ford, T. C. and Rickwood, D., Biol. Cell **45** (1982) 316.

Interaction of a DNA-binding protein, the gene product of D5 of bacteriophage T5, with double-stranded DNA: analysis by metrizamide gradient centrifugation. Fujimura, R. K. and Roop, B. C., J. Biol. Chem. **257** (1982) 14811-14816.

Sedimentation des ribonucleoproteines cytoplasmiques et nucleaires dans des milieux iodes. Houssais, J. F., Biol. Cell **45** (1982) 478.

Separation of proteins, nucleic acids and nucleoproteins on Nycodenz, a new, nonionic density gradient medium. Rickwood, D. and Ford, T. C., Biochem. Soc. Trans. **10** (1982) 363-364.

Separation of macromolecules and macromolecular complexes. Rickwood, D. and Ford, T. C., Hoppe-Seyler's Z. Physiol. Chem. **363** (1982) 1001.

Buoyant densities of macromolecules, macromolecular complexes and cell organelles in Nycodenz gradients. Ford, T. C., Rickwood, D. and Graham, J., Anal. Biochem. **128** (1983) 232-239.

3. Nucleoproteins

Isopycnic centrifugation of sheared chromatin in metrizamide gradients.

Rickwood, D., Hell, A. and Birnie, G. D., FEBS Lett. **33** (1973) 221-224.

Use of metrizamide for the isopycnic fractionation of unfixed ribonucleoprotein particles. Mullock, B. M. and Hinton, H., Biochem. Soc. Trans. **1** (1973) 578-581.

The use of metrizamide for the fractionation of ribonucleoprotein particles. Hinton, R. H., Mullock, B. M. and Gilhuus-Moe, C. C. in *Methodological Developments in Biochemistry,* vol. 4 (Ed. E. Reid) pp. 103-110. Longman, New York (1974).

Isopycnic centrifugation of sheared L-cell chromatin in metrizamide gradients. Monahan, J. J. and Hall, R. H. Nucleic Acids Res. **1** (1974) 1359-1370.

Use of metrizamide to separate cytoplasmic ribonucleoprotein particles in muscle cell cultures: method for the isolation of messenger RNA independent of its poly(A) content. Buckingham, M. E., Gros, F., FEBS Lett. **53** (1975) 355-359.

Separation of ribonucleoprotein (hn-RNP) particles from L cell nuclei into two classes, using isopycnic centrifugation in metrizamide gradients. Houssais, J.F., FEBS Lett. **56** (1975) 341-347.

Metrizamide density gradients of sea urchin chromatin: fractions rich and poor in nascent DNA. Levy, A., Jakob, K. M. and Moav, B., Nucleic Acids Res. **2** (1975) 2299-2303.

Preparation of chromatin from tissue culture cells, a convenient method. Monahan, J. J. and Hall, R. H., Anal. Biochem. **65** (1975) 187-203.

SV40, nucleoprotein complexes, structural modifications after isopysnic centrifugation in metrizamide gradients. Pignatti, P. F., Cremisi, C., Croissant, O. and Yaniv, M., FEBS Lett. **60** (1975) 369-373.

Appearance of rapidly labelled, high molecular weight RNA in nuclear ribonucleoprotein: release from chromatin and association with protein. Augenlicht, L. H. and Lipkin, M., J. Biol. Chem. **251** (1976) 2592-2599.

Isopycnic centrifugation of ribosomal precursor particles in metrizamide gradients. Auger, M.-A. and Longuet, M., Comptes Rendus Hebdomadaires des Seances de l'Academie des Sciences Serie D, Sciences Naturelles (France) **282** (1976) 1889-1892.

The separation of cytoplasmic ribonucleoprotein particles by density-gradient centrifugation in metrizamide. Buckingham, M. E. and Gros, F. in *Biological Separations in Iodinated Density-gradient Media* (Ed. Rickwood, D.) pp. 71-80. IRL Press Ltd., Oxford and Washington (1976).

The use of metrizamide density-gradients to obtain a chromatin component from mammalian cells enriched in DNA-replicating regions. Burke, F. and Pearson, K., Biochem. Soc. Trans. **4** (1976) 787-789.

The use of metrizamide as a density-gradient medium in studies of rat-liver polysomes. Dissous, C., Lempereur, C., Verwaerde, C. and Krembel, J., Eur. J. Biochem. **64** (1976) 361-367.

The use of metrizamide density gradients to separate heterogeneous nuclear ribonucleoprotein particles of mammalian cells into two fractions. Houssais, J. F. in *Biological Separations in Iodinated Density-gradient Media* (Ed. D. Rickwood) pp. 81-87. IRL Press Ltd., Oxford and Washington (1976).

Evidence for an association of most nuclear RNA with chromatin. Kimmel, C. B., Sessions, S. K. and MacLeod, M. C., J. Mol. Biol. **102** (1976) 177-191.

Separation of sea urchin chromatin by metrizamide into fractions which are enriched or depleted in nascent DNA. Levy, A., Jakob, K. M. and Moav, B. in *Biological Separations in Iodinated Density-gradient Media* (Ed. D. Rickwood) pp. 51-55. IRL Press Ltd., Oxford and Washington (1976).

Chromatin fractionation using metrizamide gradients. Malcolm, S. and Paul, J., in *Biological Separations in Iodinated Density-gradient Media* (Ed. D. Rickwood) pp. 41-49. IRL Press Ltd., Oxford and Washington (1976).

Distribution of histones in alkali-denatured chromatin, studied by isopycnic centrifugation in alkaline metrizamide gradients. Russev, G. and Tsanev, R., Nucleic Acids Res. **3** (1976) 697-707.

Isopycnic centrifugation of mammalian metaphase chromosomes. Wray, W. in *Biological Separations in Iodinated Density-gradient Media* (Ed. D. Rickwood) pp. 57-69. IRL Press Ltd., Oxford and Washington (1976).

Isopycnic centrifugation of metaphase chromosomes in metrizamide. Wray, W., FEBS Lett. **62** (1976) 202-207.

Isopycnic centrifugation of metaphase chromosomes using iodinated density gradient media. Wray, W., J. Cell Biol. **70** (1976) 257a.

Interaction of chromosomal proteins with BrdU substituted DNA as determined by chromatin-DNA competition. Bick, M. D. and Devine, E. A., Nucleic Acids Res. **4** (1977) 3687-3700.

Characterisation of an endogenous protein kinase activity in ribonucleoprotein structures containing heterogeneous nuclear RNA in HeLa cell nuclei. Blanchard, J. M., Brunel, C. and Jeanteur, P., Eur. J., Biochem. **79** (1977) 117-131.

Metrizamide dissociates nuclear particles containing heterogeneous nuclear RNA. Gattoni, R., Stevenin, J. and Jacob, M., Nucleic Acids Res. **4** (1977) 3931-3941.

Nuclear protein transitions in rat-testis spermatids. Grimes, S. R., Meistrich, M. L., Platz, R. D. and Hnilica, L. S., Exp. Cell Res. **110** (1977) 31-39.

Isolation of ribosomal gene chromatin. Higashinakagawa, T., Wahn, J. and

Reeder, R. H., Develop. Biol **55** (1977) 375-386.

A comparative study on the two classes of heterogeneous nuclear ribonucleoprotein particles separated in metrizamide density gradients, by electrophoresis of proteins and chase experiments: evidence for two distinct subfractions of hnRNP in mammalian nuclei. Houssais, J. F., Mol. Biol. Rep. **3** (1977) 251-261.

Use of metrizamide for the separation of chromosomes. Wray, W., Methods in Cell Biol. **15** (1977) 111-125.

Fractionation of chromatin by buoyant density gradient sedimentation in metrizamide. Birnie, G. D. in *Methods in Cell Biology,* vol. 18 (Eds Stein, G., Stein, J. and Kleinsmith, L. J.) pp. 23-39. Academic Press Inc., New York (1978).

Purification of yeast chromatin in metrizamide gradients. Gullov, K. B., Friis, J., Exp. Mycol. **2** (1978) 161-172.

A general method for the isolation and partial characterisation of mitochondrial nucleoids by centrifugation in metrizamide gradients. Rickwood, D. and Jurd, R. D., Biochem. Soc. Trans. **6** (1978) 266-268.

Fractionation of unfixed chromatin by buoyant density centrifugation in gradients containing 3-iodo-1,2 propanediol and metrizamide. Senshu, T. and Ohashi, M., J. Biochem. (Tokio) **83** (1978) 639-646.

Composition and stability of sea urchin mitochondrial nucleoids. Sevaljevic, L., Petrovic, L. S. and Rickwood, D., Mol. Cell Biochem. **21** (1978) 139-143.

Comparison of susceptibility to staphylococcal nuclease and behaviour in metrizamide gradients, of normal and 5-bromodeoxyuridine-substituted chromatin from *Physarum polycephalum*. Tyniec, B., Staron, K., Jerzmanowski, A. and Toczko, K., Acta Biochim. Pol. **25** (1978) 273-280.

Effect of salts and chromatin concentrations on the buoyant density of chromatin in metrizamide gradients. Kondo, T., Nakajima, Y. and Kawakami, M., Biochim. Biophys. Acta, **561** (1979) 526-534.

Comparative characterisation of mitochondrial nucleoids and of nuclear chromatin of sea urchin embryos. Sevaljevic, L., Rickwood, D. and Tomovic, M., Mol. Cell Biochem. **23** (1979) 53-62.

Isolation of myosin messenger ribonucleoprotein particles which contain a protein fraction affecting myosin synthesis. Bester, A. J., Durrheim, G., Kennedy, D. S. and Heywood, S. M., Biochem. Biophys. Res. Commun. **92** (1980) 524-531.

Characterisation of protein kinase activity associated with rat-liver polysomal messenger ribonucleoprotein particles. Cardelli, J. and Pitot, H. C., Biochemistry (USA), **19** (1980) 3164-3169.

Isolation of newly replicated chromatin by using shallow metrizamide gradients. Murphy, R. F., Wallace, R. B. and Bonner, J., Proc. Natl. Acad. Sci. (USA) **77** (1980) 3336-3340.

Nuclear ribonucleoprotein particles from adenovirus infected HeLa cells. Blanchard, J. M. and Weber, J. Mol. Biol. Rep. **7** (1981) 107-113.

Properties of ribosomes centrifuged on metrizamide gradients. Rickwood, D. and Jones, C., Biochim. Biophys. Acta, **654** (1981) 26-30.

Purification of messenger ribonucleoprotein particles from rabbit reticulocytes by zonal centrifugation in metrizamide. Rittschof, D. and Traugh, J. A., Eur. J. Biochem. **115** (1981) 45-52.

Sedimentation des ribonucleoproteines cytoplasmiques et nucleaires dans des milieux iodes. Houssais, J. F., Biol. Cell **45** (1982) 478.

Metrizamide gradient centrifugation of histone-DNA complexes. Rzeszowska-Wolny, J., Grobner, S., Filipski, J. and Chorazy, M., Acta Biochim. Polonica **29** (1982) 289-298.

4. **Membranes and Cell Organelles**

The uses of metrizamide in the fractionation of nuclei from brain and liver tissue, by zonal centrifugation. Mathias, A. P. and Wynter, C. V. A., FEBS Lett. **33** (1973) 18-22.

Isopycnic centrifugation of rat-liver subcellular particles in sucrose and in metrizamide. Collot, M., Wattiaux-de Coninck, S. and Wattiaux, R. in *Biological Separations in Iodinated Density-gradient Media* (Ed. D. Rickwood) pp. 89-96. IRL Press Ltd., Oxford and Washington (1976).

Centrifugation of liver microsomes on metrizamide density gradients. Fehrnstrom, H., Eriksson, L. C. and Dallner, G., Prep. Biochem. **6** (1976) 133-145.

Isolation of vacuoles from root storage tissue of *Beta vulgaris*. Leigh, R. A. and Branton, D., Plant Physiol. **58** (1976) 656-662.

Applications of a metrizamide gradient to the purification of glycogen particles. Guenard, D., Morange, M. and Buc, H., FEBS Lett. **76** (1977) 262-265.

Some physical properties of adrenal medulla chromaffin granules, isolated by a new continuous iso-osmotic density gradient method. Morris, S. J. and Schovanka, I., Biochim. Biophys. Acta, **464** (1977) 53-64.

Separation of isolated cell nuclei of tissues of the meal moth *(Ephestia kuehniella)* by metrizamide gradient centrifugation. Schmidt, E. R., Differentiation, **7** (1977) 107-112.

Isolation of a purified preparation of rat-liver lysosomes by centrifugation on a metrizamide gradient. Wattiaux, R., Wattiaux-de Coninck, S. and Dubois,

F., Arch. Internat. Physiol. Biochim. **85** (1977) 1033-1034.

The separation of *Xenopus laevis* nuclei from melanosomes and other cytoplasmic components. Gambino, J., Risley, M. S. and Eckhardt, R. A., J. Cell Biol. **79** (1978) 253a.

Isolation of amplified nucleoli from Xenopus oocytes. Wahn, H. L., Reeder, R. H. and Higashinakagawa, T., Methods in Cell Biol. **18** (1978) 55-60.

Isolation of rat-liver lysosomes by isopycnic centrifugation on a metrizamide gradient. Wattiaux, R., Wattiaux-de Coninck, S., Ronveaux-Duval, M. F. and Dubois, F., J. Cell Biol. **78** (1978) 349-368.

The isolation and lipid composition of subcellular fractions derived from neuronal perikarya isolated in bulk from rabbit cerebral cortex. Baker, R. R., Brain Research, **169** (1979) 65-82.

Subcellular fractionation of brown adipose tissue. Giacobino, J. P., J. Supramol. Struct. **11** (1979) 445-449.

Tissue culture cell fractionation, fractionation of cellular membranes from iodine-125-lactoperoxidase labelled Lettree cells, homogenised by bicarbonate induced lysis. Resolution of membranes by zonal centrifugation and in sucrose and metrizamide gradients. Graham, J. M. and Coffey, K. H. M., Biochem. J. **182** (1979) 173-180.

Isolation of nucleoli from *Tetrahymena pyriformis*. Higashinakagawa, T., Sezaki, M. and Kondo, S., Develop. Biol. **69** (1979) 601-611.

Purification of *Xenopus laevis* embryo and liver nuclei, using metrizamide. Risley, M. S., Gambino, J. and Eckhardt, R. A., Develop. Biol. **68** (1979) 299-303.

A method for the isolation of alpha granules from human platelets. Gogstad, O., Thromb. Res. **20** (1980) 669-681.

The mitochondrial heterogeneity in bull frog *(Rana catesbeiana)* liver. Goto, Y., Makino, Y., Shimizu, J. and Shukuya, R., Comp. Biochem. and Physiol. B Comp. Biochem. **66** (1980) 609-610.

Isolation of renin granules from rat-kidney cortex by isotonic or hyperosmotic metrizamide-sucrose gradients. Manisto, P. T. and Poisner, A. M., Prep. Biochem. **10** (1980) 297-316.

Isolation and characterisation of autophagic lysosomes from rat-liver. Marzella, L., Ahlberg, J. and Glaumann, H., Eur. J. Cell Biol. **22** (1980) 199.

Metrizamide isopycnic centrifugation for the isolation of macronuclei and micronuclei from *Paramecium bursaria* and *Paramecium trichium*. Shiomi, Y., Higashinakagawa, T., Saiga, H., and Mita, T., J. UOEH, **2** (1980) 323-330.

Isolation of rat somatotroph and mammotroph secretory granules by equilibrium density centrifugation in a linear metrizamide gradient. Slaby, F. and Farquhar, M. G., Mol. Cell. Endocrinol. **18** (1980) 21-32.

Isolation and initial characterisation of a lymphocyte cap structure. Bourguignon, G. J. and Bourguignon, L. Y., Biochim. Biophys. Acta, **646** (1981) 109-118.

Characterisation of gut hormone storage granules from normal human jejunum, using metrizamide density gradients. Dawson, J., Bryant, M. G., Bloom, S. R. and Peters, T. J., Regul. Pept. **2** (1981) 305-315.

The isolation of purified neurosecretory vesicles from bovine neurohypophysis, using iso-osmolar density gradients. Russell, J. T., Anal. Biochem. **113** (1981) 229-238.

Formation of hybrid nucleosomes containing new and old histones. Russev, G. and Hancock, R., Nucleic Acids Res. **9** (1981) 4129-4137.

Purification of vacuoles from *Neurospora crassa*. Vaughn, L. E. and Davis, R. H., Mol. Cell. Biol. **1** (1981) 797-806.

The isolation and purification of mouse epidermal nuclei. Wortman, M. J. and Segal, A., Anal. Biochem. **115** (1981) 147-150.

Isolation of peroxisomes by isopycnic centrifugation in metrizamide gradients and analysis of the distribution of constituents with multiple localisation by restrained linear regression. Bronfman, M., Leighton, F. and Feytmans, E., Ann. N. Y. Acad. Sci. **386** (1982) 408-410.

Characterisation of the membrane proteins of rat-liver lysosomes; composition, enzyme activities and turnover. Burnside, J. and Schneider, D. L., Biochem. J. **204** (1982) 525-534.

Fractionation of mammalian cell membranes in iodinated density gradient media. Graham, J., Biol. Cell **45** (1982) 476.

Subfractionation of rat-liver endoplasmic reticulum in iodinated density gradient media. Graham, J. and Wagner, S., Hoppe-Seyler's Z. Physiol. Chem. **363** (1982) 1000.

Glycerolipid biosynthesis in peroxisomes via the acyl dihydroxyacetone pathway. Hajra, A. K. and Bishop, J. E., Ann. N. Y. Acad. Sci. **368** (1982) 170-182.

Subcellular distribution of particle associated enzymes in horse neutrophil leukocytes. Heynemann, R. A. and Vercauteren, R. E., Enzyme **27** (1982) 141-148.

Isolation of autophagic vacuoles from rat liver. Morphological and biochemical characterisation. Marzella, L., Ahlberg, J. and Glaumann, H., J. Cell Biol. **93** (1982) 144-154.

Biochemical and biophysical characterisation of insulin granules isolated from rat pancreatic islets by an iso-osmotic gradient. Mathews, E. K., McKay, D. B., O'Connor, M. D. L. and Borowitz, J. L., Biochem. Biophys. Acta, **715** (1982) 80-89.

Isolation of rat-liver peroxisomes by vertical rotor centrifugation using self-generating cpd545 (Nycodenz) density gradients. Osmundsen, H. and Cervenka, J., Hoppe-Seyler's Z. Physiol. Chem. **363** (1982) 1000.

Initial characterisation of human platelet mepacrine-labelled granules, isolated using a short metrizamide gradient. Rendu, F., Lebret, M., Nurden, A. T. and Caen, J. P., Brit. J. Haematol. **52** (1982) 241-251.

Isolation of subcellular organelles in sucrose, metrizamide and cpd 545 (Nycodenz) density gradients. Wattiaux, R., Wattiaux-de Coninck, S. and Vandenberghe, A., Biol. Cell **45** (1982) 475.

Behaviour of subcellular organelles from rat-liver, kidney and lung, in metrizamide density gradients. Winter, R., Sutterlin, U., Gluck, M. and Seidel, A., Biol. Cell **45** (1982) 316.

5. **Cells**

Use of metrizamide as a gradient medium for the isopycnic separation of rat-liver cells. Munthe-Kaas, A. C. and Seglen, P. O., FEBS Lett. **43** (1974) 252-256.

Fractionation of *Xenopus laevis* testicular cells by equilibrium density gradient centrifugation in metrizamide. Risley, M. S., Eckhardt, R. A. and Eppig, J. J., J. Cell Biol. **67** (1975) 363a.

Platelet isolation and function Part 1, from platelet-rich plasma. Comparison of two methods, gel-filtration and centrifugation on an albumin gradient. Part 2, New method from total blood centrifugation on metrizamide gradients. Levy-Toledano, S., Bredoux, R., Rendu, F., Jeanneau, C., Savariau, E. and Dassin, E., Nouvelle Revue Francaise d'Hemotologie **16** (1976) 367-380.

The use of metrizamide for the separation of rat-liver cells. Seglen, P. O. in *Biological Separations in Iodinated Density-gradient Media* (Ed. D. Rickwood) pp. 107-121. IRL Press Ltd., Oxford and Washington (1976).

Isolation of highly purified Leydig cells by density gradient centrifugation. Conn, P. M., Tsurahara, T., Dufau, M. and Catt, K. J., Clin. Chim. Acta, **79** (1977) 639-642.

Resolution of aortic cell populations by metrizamide density gradient centrifugation. Haley, N. J., Shio, H. and Fowler, S., Lab. Invest. (US) **37** (1977) 287-296.

Isolation and characterisation of Kupffer and endothelial cells from rat liver. Knook, D. L., Blansjaar, N. and Sleyster, E. C., Exp. Cell Res. **109** (1977) 317-329.

Quantitative analysis of DNA, RNA, protein and glycogen in isolated rat hepatocytes, separated by metrizamide density gradients. Mayer, D., Stoehr, M. and Lange, L., Cytobiologie, **15** (1977) 321-334.

Isolation of functionally intact pancreatic islets by centrifugation in metrizamide gradients. Raydt, G., Hoppe-Seyler's Z. Physiol. Chem. **358** (1977) 1369-1373.

Differentiation of *Xenopus laevis* spermatogenic cells *in vitro*. Risley, M. S. and Eckhardt, R. A., J. Cell Biol. **79** (1978) 181a.

Separation of erythrocytes infected with murine malaria parasites in metrizamide gradients. Eugui, E. M. and Allison, A. C., Parasitology, **79** (1979) 267-276.

Metrizamide density gradients for the separation of different developmental stages of malarial parasites. Eugui, E. M. and Allison, A. C., Bul. WHO **57** (1979) Suppl. 1, 181-187.

Human platelet isolation from whole blood, on metrizamide gradients. Levy-Toledano, S., Rendu, F., Bredoux, R., de la Baume, H., Dmozynska, A., Savariau, E. and Caen, J. P., in *Separation of Cells and Subcellular Elements*. pp. 83-86, Pergamon Press, Oxford, England (1979).

Dissociation and separation of *Xenopus laevis* spermatogenic cells. Risley, M. S. and Eckhardt, R. A., J. Expl. Zoo. **207** (1979) 93-106.

Evidence for increased proteolytic activity in aging human fibroblasts. Shakespeare, V. A. and Buchanan, J. H., Gerontology, **25** (1979) 305-313.

Metrizamide gradient purification of mouse tumour cells. Stackpole, C.W., Cremona, P. and Kassel, R. L., J. Immunol. Methods, **30** (1979) 231-243.

Unique distribution of glycoprotein receptors on parenchymal and sinusoidal cells of rat liver. Steer, C. and Clarenburg, R., J. Biol. Chem. **254** (1979) 4457-4461.

Isolation of ACTH responsive cells from rat adrenal cortex and the determination of the density of male and female cells. Ungar, F., Hsiao, J., Greene, J. M. and Headon, D. R., FEBS Lett. **108** (1979) 331-334.

Parenchymal cells purified from Xenopus liver and maintained in primary culture, synthesize vitellogenin in response to estradiol 17β and serum albumin in response to dexamethasone. Wangh, L. J., Osborne, J.A., Hentschel, C.C. and Tilly, R., Develop. Biol. **70** (1979) 479-499.

Sebaceous gland differentiation 2. The isolation, separation and characterisation of cells from mouse preputial gland. Wheatley, V. R., Potter, J. E. R. and Lew, G., J. Invest. Dermatol. **73** (1979) 291-296.

Fractionation and characterisation of rat-liver hepatocytes isolated from ethanol induced fatty liver. Kondrup, J., Bro, B., Dich, J., Grunnet, N. and

Thieden, H. I. D., Lab. Invest. **43** (1980) 182-190.

Isopycnic separation of human, pigeon and guinea-pig erythrocytes. Ogunmola, G. B., Dada, O. A. and Ejike, E. N., Acta Haematol. **63** (1980) 312-316.

Separation of different cell types from a transplanted hepatoma. Siow, Y.-F., Dobrota, M. and Hinton, R. H., Biochem. Soc. Trans. **8** (1980) 108-109.

Properties of DNA polymerase from purified nuclei and DNA synthesizing complexes of human spermatozoa. Witkin, S. S., J. Reprod. Fertil. **59** (1980) 409-420.

Partial separation and biochemical characteristics of periportal and perivenous hepatocytes from rat liver. Bengtsson, B. G., Kiessling, K.-H., Smith-Kielland, A., and Morland, J., Eur. J. Biochem. **118** (1981) 591-598.

Separation of rat-liver cells by means of Nycodenz. Berg, T., Holte, K., Naess, L., Eskild, W. and Blomhoff, R., Biol. Cell **45** (1982) 477.

Separation of mononuclear cells, granulocytes and monocytes with Nycodenz, a new gradient medium. Boyum, A., Hoppe-Seyler's Z. Physiol. Chem. **363** (1982) 999.

Effects of estradiol administration *in vivo* on testosterone production, in two populations of rat Leydig cells. Keel, B. A. and Abney, T. O., Biochem. Biophys. Res. Comm. **107** (1982) 1340-1348.

The fat storing cells of rat liver: their isolation and purification. Knook, D. L., Seffelaar, A. M. and de Leeuw, A. M., Expl. Cell Res. **139** (1982) 468-471.

Isolation of parenchymal cell derived particles from non-parenchymal rat-liver cell fractions. Nagelkerke, J. F., Barto, K. P. and Van Berkel, T. J. C., Expl. Cell Res. **138** (1982) 183-191.

Biochemical characterisation of isolated rat retinal cells: the gamma-aminobutyric acid system. Schaeffer, J. M., Exp. Eye Res. **34** (1982) 715-727.

Protein synthesis in different populations of rat hepatocytes, separated according to density. Smith-Kielland, A., Bengtsson, G., Svendsen, L. and Morland, J., J. Cell Physiol. **110** (1982) 262-266.

Wet and dry bacterial spore densities, determined by buoyant sedimentation. Tisa, L. S., Koshikawa, T. and Gerhardt, P., Appl. Environ. Microbiol. **43** (1982) 1307-1310.

Isopycnic centrifugation of chromosomes and cells on Nycodenz gradients. Wray, W., Johnson, J., Gollin, S. M., Barr, C. and Wray, V. P., J. Cell Biol. **95** (1982) 465a.

6. **Viruses**

SV 40 nucleoprotein complexes: structural modifications after isopycnic centrifugation in metrizamide gradients. Pignatti, P. F., Cremisi, C., Croissant, O. and Yaniv, M., FEBS Lett. **60** (1975) 369-373.

Purification of virions and nucleocapsids of Herpes simplex virus, by means of metrizamide and sodium metrizoate gradients. Blomberg, J., Bjorck, E., Olofsson, S., Berg, G. and Lycke, E., Arch. Virol. **50** (1976) 271-278.

Chromatin-like structures in polyoma virus and Simian virus 40 lytic cycle. Cremisi, C., Pignatti, P. F., Croissant, O. and Yaniv, M., J. Virol. **17** (1976) 204-211.

Subfractionation of caesium chloride-purified H-1 parvovirus on metrizamide gradients. Kongsvik, J. R., Hopkins, M. S. and Ellem, K. A. O., Virology, **96** (1976) 646-651.

The influence of a tri-iodinated benzamido derivative of glucose on measles virus. Vanden Berghe, D. A. R., Arch. Int. Physiol. Biochim. **84** (1976) 415-416.

The inactivation of polio, Coxsackie and Semliki Forest viruses by a tri-iodinated derivative of glucose. Vanden Berghe, D. A. R., Van der Groen, G. and Pattyn, S. R., Arch. Int. Physiol. Biochim., **84** (1976) 670-671.

The effect of metrizamide on the multiplication of animal RNA viruses in mouse organ cultures. Van der Groen, G., Vanden Berghe, D. A. R., Delgadillo, R. A. and Pattyn, S. R., Arch. Int. Physiol. Biochim. **84** (1976) 671-672.

Metrizamide and the infectivity of animal RNA viruses. Vanden Berghe, D. A. R., Van der Groen, G. and Pattyn, S. R. in *Biological Separations in Iodinated Density-gradient Media* (Ed. D. Rickwood) pp. 175-183. IRL Press Ltd., Oxford and Washington (1976).

Separation of viruses in metrizamide density gradients. Wunner, W. H., Buller, R. M. and Pringle, C. R. in *Biological Separations in Iodinated Density-gradient Media* (Ed. D. Rickwood) pp. 159-173. IRL Press Ltd., Oxford and Washington (1976).

Respiratory syncytial pneumonia virus proteins. Wunner, W. H. and Pringle, C. R., Virology, **73** (1976) 228-243.

Purification of polyhedra of a cytoplasmic polyhedrosis virus from soil, using metrizamide. Hukahara, T., J. Invert. Pathol. **30** (1977) 270-272.

Metrizamide, a new reagent for purification of RNA oncogenic viruses associated with bovine enzootic leukemia and human leukemias. Portetelle, D., Ghysdael, J., Burny, A., Dekegel, D., Mammerickx, M., Prevost, J. M. and Chantrenne, H., Arch. Int. Physiol. Biochim. **85** (1977) 194-195.

Effect of different gradient solutions on the buoyant density of scrapie infectivity. Brown, P., Green, E. N. and Gajausek, D. C., Proc. Soc. Exp. Biol. Med. **158** (1978) 513-516.

Purification of mengovirus by Freon extraction and chromatography on protein coated controlled pore glass. Gschwender, H. H. and Traub, P., Arch. Virol. **96** (1978) 327-336.

Early events in influenza virus infection: the formation of a nucleoplasmic ribonucleoprotein complex from the input virion. Hudson, J. B., Flawith, J. and Dimmock, N. J., Virology **86** (1978) 167-176.

Structural analysis of viral replicative intermediates, isolated from adenovirus Type 2 infected HeLa cell nuclei. Kedinger, C., Brison, O., Perrin, F. and Wilhelm, J., J. Virol. **26** (1978) 364-379.

Purification of channel catfish virus, a fish herpes virus. Robin, J. and Rodrigue, A., Can. J. Microbiol. **24** (1978) 1335-1338.

Separation of complete core of hepatitis B Dane particles by ultracentrifugation in metrizamide gradients. Takahashi, T. in Japan Medical Research Foundation Publication No. 6; Hepatitis viruses. pp. 75-83. University Park Press, Baltimore, USA (1978).

Characterisation of murine pneumonia virus proteins. Cash, P., Preston, C. M. and Pringle, C. R., Virology, **96** (1979) 442-452.

Equilibrium centrifugation of measles virus. Huybreghts, G., Vanden Berghe, D. A. and Pattyn, S. R., Biochem. Biophys. Res. Commun. **93** (1980) 999-1004.

A metrizamide impermeable capsid in the DNA packaging pathway of bacteriophage T7. Serwer, P., J. Mol. Biol. **138** (1980) 65-91.

Isolation of Dane particles containing a DNA strand, by metrizamide density gradient. Takahashi, T., Kaga, K., Akahane, Y., Yamashita, T., Miyakawa, Y. and Mayumi, M., J. Med. Microbiol. **13** (1980) 163-166.

Isolation of influenza virus ribonucleoproteins by isopycnic centrifugation on metrizamide gradients. Twigden, S. J. and Horisberger, M. A., Experientia, **36** (1980) 1449.

Lipid composition for two mycoplasmaviruses, MV-Lg-L172 and MV L2. Al-Shammari, A. J. N. and Smith, P. F., J. Gen. Virol. **54** (1981) 455-458.

Purification of infectious rubella virus by gel filtration on Sepharose 2B compared to gradient centrifugation in sucrose-sodium metrizoate and metrizamide. Trudel, M., Marchessault, F. and Payment, P., J. Virol. Methods, **2** (1981) 141-148.

Interactions between phage λ replication proteins, λ DNA and minicell membrane. Zyliez, M. and Taylor, K., Eur. J. Biochem. **113** (1981) 303-309.

Nonionic density gradient purification of Epstein-Barr virus and identification of a virus associated protein kinase. Fowler, E., Hutt-Fletcher, C. M. and Feighny, R. J., J. Cell Biol. **95** (1982) 460a.

Fractionation of Sendai virus in iodinated density gradient media. Graham, J. and Micklem, K., Biol. Cell **45** (1982) 316.

New density gradient media and animal RNA viruses. Vanden Berghe, D. A. R. and Huybreghts, G., Biol. Cell, **45** (1982) 474.

INDEX

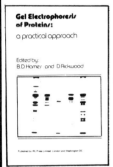

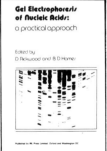